WITHDRAWN
D0275531

CHARACTERIZATION OF OPTICAL MATERIALS

MATERIALS CHARACTERIZATION SERIES
Surfaces, Interfaces, Thin Films

Series Editors: C. Richard Brundle and Charles A. Evans, Jr.

Series Titles

Encyclopedia of Materials Characterization, C. Richard Brundle, Charles A. Evans, Jr., and Shaun Wilson

Characterization of Metals and Alloys, Paul H. Holloway and P. N. Vaidyanathan

Characterization of Ceramics, Ronald E. Loehman

Characterization of Polymers, Ned J. Chou, Steven P. Kowalczyk, Ravi Saraf, and Ho-Ming Tong

Characterization in Silicon Processing, Yale Strausser

Characterization in Compound Semiconductor Processing, Yale Strausser

Characterization of Integrated Circuit Packaging Materials, Thomas M. Moore and Robert G. McKenna

Characterization of Catalytic Materials, Israel E. Wachs

Characterization of Composite Materials, Hatsuo Ishida

Characterization of Optical Materials, Gregory J. Exarhos

Characterization of Tribological Materials, William A. Glaeser

Characterization of Organic Thin Films, Abraham Ulman

CHARACTERIZATION OF OPTICAL MATERIALS

EDITOR

Gregory J. Exarhos

MANAGING EDITOR

Lee E. Fitzpatrick

BUTTERWORTH-HEINEMANN
Boston London Oxford Singapore Sydney Toronto Wellington

MANNING
Greenwich

M

This book was acquired, developed, and produced by Manning Publications Co.
Design: Christopher Simon
Copyediting: Deborah Oliver
Typesetting: Stephen Brill

∞ Recognizing the importance of preserving what has been written, it is the policy of Butterworth-Heinemann and of Manning to have the books they publish printed on acid-free paper, and we exert our best efforts to that end.

Library of Congress Cataloging–in–Publication Data
Characterization of optical materials/editor, Gregory J. Exarhos.
p. cm.—(Materials characterization series)
Includes bibliographical references and index.
ISBN 0-7506-9298-7
1. Optical materials. 2. Optical materials—Surfaces. 3. Optical materials—Design and construction. I. Exarhos, Gregory J. II. Series.
QC374.C48 1992 92–39023
621.36—dc20 CIP

Butterworth-Heinemann
80 Montvale Avenue
Stoneham, MA 02180

Manning Publications Co.
3 Lewis Street
Greenwich, CT 06830

10 9 8 7 6 5 4 3 2 1

Printed in the United States of America

Contents

Preface to Series

This Materials Characterization Series attempts to address the needs of the practical materials user, with an emphasis on the newer areas of surface, interface, and thin film microcharacterization. The Series is composed of the leading volume, *Encyclopedia of Materials Characterization*, and a set of about 10 subsequent volumes concentrating on characterization of individual materials classes.

In the *Encyclopedia*, 50 brief articles (each 10 to 18 pages in length) are presented in a standard format designed for ease of reader access, with straightforward technique descriptions and examples of their practical use. In addition to the articles, there are one-page summaries for every technique, introductory summaries to groupings of related techniques, a complete glossary of acronyms, and a tabular comparison of the major features of all 50 techniques.

The 10 volumes in the Series on characterization of particular materials classes include volumes on silicon processing, metals and alloys, catalytic materials, integrated circuit packaging, etc. Characterization is approached from the materials user's point of view. Thus, in general, the format is based on properties, processing steps, materials classification, etc., rather than on a technique. The emphasis of all volumes is on surfaces, interfaces, and thin films, but the emphasis varies depending on the relative importance of these areas for the materials class concerned. Appendixes in each volume reproduce the relevant one-page summaries from the *Encyclopedia* and provide longer summaries for any techniques referred to that are not covered in the *Encyclopedia*.

The concept for the Series came from discussion with Marjan Bace of Manning Publications Company. A gap exists between the way materials characterization is often presented and the needs of a large segment of the audience—the materials user, process engineer, manager, or student. In our experience, when, at the end of talks or courses on analytical techniques, a question is asked on how a particular material (or processing) characterization problem can be addressed the answer often is that the speaker is "an expert on the technique, not the materials aspects, and does not have experience with that particular situation." This Series is an attempt to bridge this gap by approaching characterization problems from the side of the materials user rather than from that of the analytical techniques expert.

We would like to thank Marjan Bace for putting forward the original concept, Shaun Wilson of Charles Evans and Associates and Yale Strausser of Surface Science Laboratories for help in further defining the Series, and the Editors of all the individual volumes for their efforts to produce practical, materials user based volumes.

C. R. Brundle *C. A. Evans, Jr.*

Preface

The design and manufacture of advanced optical materials have been driven by a multidisciplinary approach from which new components and integrated optical devices have evolved. The diversity of optical materials—including metals, polymers, glasses, ceramics, semiconductors, and composites—provides a challenge to the analyst charged with characterizing optical surfaces and interfaces and developing associated structure–property relationships. Analysis of most optical materials must include not only measurements of optical properties, but a determination of the fundamental surface and interfacial material properties as well. For example, correlating the optical response of a material with its microstructure, residual interfacial stress, phase purity, and surface-roughness can lead to a refinement of processing methods in order to secure the optimum material for a particular application.

This volume—one component of the *Materials Characterization Series: Surfaces, Interfaces, Thin Films*—focuses on the kind of information derived from the principal analytical methods currently used to characterize optical materials. This information is useful for identifying the key parameters that control the optical response of a material. The theory and methodology of the analytical methods used for the surface and interfacial characterization of optical materials are discussed in the lead volume of the series, *Encyclopedia of Materials Characterization*, and are summarized in the appendix of this book. The intent of this volume is not to discuss instrumental methods in great detail, but to provide the necessary background information to permit a constructive dialogue to be initiated between the researcher with a specific problem to address and the technical specialist skilled in applying a particular method.

This book consists of two major sections, an introductory chapter, and an extensive appendix, which summarizes the analytical methods pertinent to the characterization of optical materials. The book covers both crystalline and amorphous materials—with applications from the far infrared (~1 mm wavelength) to the vacuum ultraviolet (~100 nm wavelength) regions of the spectrum—and presents a succinct discussion of the type of information obtainable using various key surface characterization methods. The extensive list of references for each chapter may be consulted to gain a more thorough understanding of a particular subject area.

The topics covered and organization of this volume grew from a number of discussions with Marjan Bace of Manning Publications Company, who persuaded me to consider editing a text on the surface and interfacial characterization of optical materials. His direction and advice were greatly appreciated. I wish to express my gratitude to all contributing authors, who took time from their busy schedules to participate in writing this volume. I hope the readers will find our efforts of practical value. Our Managing Editor, Lee Fitzpatrick, also of Manning Publications Company, was instrumental in persuading the authors to submit their chapters in a timely manner so that the production schedule could be met. Her efforts in dealing with all involved were invaluable. Finally, I wish to acknowledge the Pacific Northwest Laboratory, operated for the U.S. Department of Energy by Battelle Memorial Institute under contract DE-AC06-76RLO 1830, for allowing me the time and resources to complete this project.

Gregory J. Exarhos

Contributors

Michael F. Becker University of Texas-Austin Austin, TX	Laser-Induced Damage to Optical Materials
Jean M. Bennett Naval Air Warfare Center China Lake, CA	Characterization of Surface Roughness
Gregory J. Exarhos Battelle Pacific Northwest Laboratory Richland, WA	Introduction
E. N. Farabaugh NIST Gaithersburg, MD	Diamond As an Optical Material
Albert Feldman NIST Gaithersburg, MD	Diamond As an Optical Material
Trevor P. Humphreys North Carolina State University Raleigh, NC	The Composition, Stoichiometry, and Related Microstructure of Optical Materials
Gerald E. Jellison, Jr. Oak Ridge National Laboratory Oak Ridge, TN	Characterization of the Near-Surface Region Using Polarization-Sensitive Optical Techniques
Peter M. Martin Battelle Pacific Northwest Laboratory Richland, WA	Multilayer Optical Coatings
Carl J. McHargue The University of Tennessee-Knoxville Knoxville, TN	Surface Modification of Optical Materials
Robert J. Nemanich North Carolina State University Raleigh, NC	The Composition, Stoichiometry, and Related Microstructure of Optical Materials
Bradley J. Pond S. Systems Corporation Albuquerque, NM	Characterization and Control of Stress in Optical Films
L. H. Robins NIST Gaithersburg, MD	Diamond As an Optical Material
D. Shechtman Technion Haifa	Diamond As an Optical Material

0

Introduction

GREGORY J. EXARHOS

The development of improved optical materials requires a thorough understanding of their surface and subsurface structure and associated chemistry so that optical properties can be modified in a controlled manner and the materials durability enhanced. Advanced surface analytical methods are used to probe these materials to discern structure–property relationships which are used for designing materials with a specific response. This volume is intended to provide the researcher an introduction to the principal methods used for the surface and interfacial analysis of materials used for optical applications.

Optical materials discussed include bulk solids, thin films, and multilayer dielectric coatings which find use in applications requiring reflection, refraction, absorption, emission, scattering, or the diffraction of infrared, visible, or ultraviolet light having wavelengths from about 100 nm to 10 mm. Included in the variety of materials covered by this definition are metals, glasses, polymers, semiconductors, ceramic oxides, carbides and nitrides, and diamond. A cadre of surface analytical techniques is required for the characterization of such a diverse collection of materials. Furthermore, methods used to analyze a specific class of optical materials may be entirely inappropriate for a different materials class. Once an appropriate analysis method or methods have been identified, the information derived from the measurements can be used in several ways, such as assuring that the optical material or device meets desired specifications and controlling interactively processing parameters during manufacture. In a more fundamental approach, empirical models that relate structural properties of the material and chemical bonding to the intrinsic optical response can be developed and refined. Such models provide a basis for optimizing processing parameters to achieve a targeted optical response. The goal of this volume is to review the most important methods for characterizing the surface and interfacial properties of optical materials and to demonstrate for each method the kind of information obtained and how to interpret it.

The surface and interfacial properties of an optical material can be probed at length scales ranging from micrometers to tenths of nanometers. Microscopic defects which control surface smoothness and coating homogeneity contribute to the scattering of light and associated degradation of the optical response. Likewise, structural perturbations at the atomic level can alter optical properties through modifications to the complex refractive index of the material. To characterize *defects* at this level, high-resolution diffraction measurements coupled with the evaluation of localized chemical bonding using molecular spectroscopic techniques may be required. Such an approach can provide information regarding the structural phase of the material, the degree of surface/interfacial stress present, and the nature of the chemical bonding associated with a particular structure. Therefore, analytical probes having sensitivity at increasing levels of spatial and depth resolution are required in order to understand the physical and chemical properties that control the associated optical response of the material. These issues are addressed in Part I of this volume where correlations between the optical response of a material, surface morphology, and associated microstructure are described.

Surface roughness is a critical parameter that can degrade the performance of optical materials through light scattering processes. In addition, modification to the optical response can result from rough surfaces in which chemical attack from the ambient environment (water) is accelerated due to the increased surface area. The assessment of surface roughness is therefore critical to the behavior and long term stability of optical materials. Surface topology is a recurrent theme throughout this volume beginning with its introduction and definition in the first chapter and continuing in later chapters, including those that focus on diamond coatings and laser damage phenomena.

A powerful technique for characterizing the optical response of materials relies on measuring the change in the properties of light reflected from a surface or interface. Ellipsometry has evolved into a powerful noninvasive surface characterization tool, and its application to studies of optical materials is reviewed in this volume. Instruments are currently available that directly interface to vacuum chambers for real-time in situ characterization during the deposition of optical coatings or modification of surfaces through ion irradiation. Data acquired using this technique can be interpreted on the basis of various empirical models (Effective Medium Approximation) designed to extract microstructural information intrinsic to the surface under investigation. Such measurements complement the electron beam microscopy techniques used to evaluate microstructure but have the advantages of being nondestructive and amenable to materials analysis in real time. Ellipsometric measurements are relatively easy to perform, but considerable effort is required to interpret the measured data. Chapter 2 provides insight into the kinds of information that can be extracted from ellipsometric measurements of optical surfaces and indicates limitations of the technique. The quality of information inferred from these optical reflection methods is strongly dependent upon the model used to interpret the data. New algorithms have been developed relating the microstructure of a material to the optical response. One example, based upon a finite element model of the microstructure in a thin film, appears in a recent article by Risser and Ferris.[1] The technique of using ellipsometric methods to provide more extensive microstructural information about surfaces and interfaces will continue to mature as the models used to interpret the data become more refined.

Surface morphology and microstructure act to perturb the optical response of materials, but the nature of chemical bonding, the deviations from stoichiometry, and the presence of impurities can influence the intrinsic optical properties to a greater extent. Standard surface probing techniques such as Auger spectroscopy (AES), X-ray photoemission spectroscopy (XPS), secondary ion mass spectroscopy (SIMS), and Rutherford backscattering spectroscopy (RBS) are very useful for the elemental identification and concentration depth profiling of thin films. However, subtle changes in chemical bonding, which can produce large changes in the optical response, often are difficult to quantify using these techniques, particularly when the material under investigation is nonconducting or when bound hydrogen is an integral component of the material.

Significant advances in the application of vibrational spectroscopy to the analysis of surfaces and interfaces have been reported during the past several years. Infrared absorption and inelastic light scattering (Raman spectroscopy) are two principal nondestructive methods that have been used extensively for the characterization of optical materials. These techniques are used routinely to analyze semiconductor materials for optoelectronic applications (Chapter 3) and multilayer dielectric films used as filters or mirrors (Chapter 5). In many cases, the optical response of the material to be examined can be used to enhance the sensitivity of the photon-based analytic methods. A paper by Friedrich and Exarhos[2] describing several Raman enhancement methods for characterizing dielectric thin films should be consulted for a more complete discussion of this subject.

The optical response of a material or multilayer coating can be modified either intentionally during processing as a means to improve performance or unintentionally during use as a result of environmental degradation. Changes in the optical properties can be traced to physical or chemical alteration of the surface in bulk materials or interfacial layers in multilayer coatings. Variations in refractive index may be achieved during processing by ion bombardment at selective energies which serves to implant impurity species at depths proportional to the ion energy, alter the surface stoichiometry at larger doses, or modify the surface microstructure. Part II of this volume deals with modifications to optical materials and the correlation of optical properties to measured perturbations in the surface and interfacial layer characteristics.

The design and development of specialty optical coatings is a principal thrust in the optical materials area: Multilayer dielectric stacks are used to regulate the transmission and reflection properties of a surface by means of interference phenomena. Variations in individual layer composition, which control refractive index and layer thickness, are used to develop antireflection (AR) and high reflection (HR) mirrors, which are used in high power laser systems and for controlling light transmission in optical windows. In many of these applications, irreversible changes in transmittance can be introduced as a result of an applied stress (mechanical, thermal, or chemical) which alters the chemical and/or physical properties of the surface. Surface analytical techniques are important for characterizing these changes in order for us to understand how a coating fails. For example, sputter-deposited coatings often are found to exhibit a columnar grain microstructure. Diffusion of water through the intergranular channels will lead to either reversible changes in the layer index, depending on the amount adsorbed, or irreversible changes, if the water chemically interacts with the dielectric material. The mechanism for this type of failure has been substantiated using a variety of surface analytical techniques. To minimize this problem, processing parameters have been modified to produce a fine-grained microstructure which retards diffusion of water and minimizes chemically reactive sites. Chapter 5 discusses the key methods used to characterize optical coatings.

Materials deposited as thin films usually exhibit some degree of interfacial stress arising in part from the difference in thermal expansion coefficient or lattice mismatch between the film and substrate material. This interfacial stress is distinguishable from the inherent stress in a thin film that varies with crystallite grain size, phase homogeneity, and the presence of nonstoichiometric phases or impurities. The magnitude of such stresses can be influenced by changing film deposition parameters or through physical or chemical interaction of the film with the environment. Both the real and imaginary parts of the refractive index change with stress, and large stresses may cause the performance of a multilayer coating to deviate from the design parameters. Chapter 6 discusses the causes of interfacial stress in optical thin films and techniques for characterizing the stress. Knowledge of the degree of stress and the stress homogeneity in these coatings is necessary for us to understand their optical response, coating–substrate adherence, and stability during use, particularly in chemically reactive environments or under high fluence irradiation.

The final chapter concerns the performance of optical surfaces, thin films, and multilayer coatings exposed to high-energy pulsed-laser irradiation. The identity and concentration of chemical impurities, surface morphology, microstructure, and thickness and stress homogeneity are important parameters which control laser damage thresholds in these materials. Photothermal methods for identifying predamage sites are introduced as relatively new surface-sensitive techniques that have significant advantages over other methods commonly used to identify likely surface regions for damage.

The appendix summarizes the key surface analytical techniques used for the surface and interfacial characterization of optical materials. Owing to the diversity of materials which comprise this category, an arsenal of possible methods is required. In addition to the methods summarized in the lead volume of this series, *Encyclopedia of Materials Characterization*, various surface sensitive microscopy techniques designed specifically for optical materials are discussed. These include total internal reflection microscopy and photothermal deflection methods.

Since optical materials belong to a relatively large number of different materials classes, the reader is encouraged to consult other volumes in the Materials Characterization Series to get a different perspective on surface analytical methods applied to a specific material. Particular volumes most relevant to the subjects presented herein are *Characterization of Metals and Alloys, Characterization of Ceramics, Characterization of Polymers, Characterization in Silicon Processing,* and *Characterization in Compound Semiconductor Processing.*

References

1 S. M. Risser and K. F. Ferris. *Materials Letters.* **14**, 99–102, 1992.

2 D. M. Friedrich and G. J. Exarhos. *Thin Solid Films.* **154**, 257–270, 1987.

Part I

Influence of Surface Morphology and Microstructure on Optical Response

1

Characterization of Surface Roughness

JEAN M. BENNETT

Contents

1.1 Introduction

Surface roughness is becoming increasingly important as optical instruments have much higher performance specifications, microelectronic circuits become more compressed with more information packed into a smaller area, and optical and magnetic disks have increased information storage density. Surface characterization techniques have now advanced to sophisticated levels to meet these needs.

Surface characterization is too large a subject to be covered adequately in a chapter of this length. Thus, only the basics will be given here, including references where more information can be obtained. In Section 1.2 we first define what we mean by surface roughness—what it is and also what it is not. Examples are given of a metal reflector and a glass surface to show how the processing of a material can affect its surface roughness. In Section 1.3 we describe how surface roughness affects optical measurements. Techniques for measuring surface roughness and scattering are given in Section 1.4. Section 1.5 contains examples of different types of surfaces and lists appropriate techniques for characterizing them. Finally, in Section 1.6 we note future directions for surface characterization and suggest places where more work is needed.

Several excellent review articles and books have been written about surface characterization. The 1982 book edited by Thomas[1] contains a detailed discussion of stylus instruments and statistics appropriate for machined surfaces. The tutorial, *Introduction to Surface Roughness and Scattering*,[2] is a simple yet comprehensive treatment of characterization of optical surfaces, including measuring instruments, surface statistics, and elementary theory. Reference 3 contains extensive references to the entire field of surface characterization and over 100 selected, previously published articles on the subject.

There are many excellent references to theoretical articles relating surface roughness to scattering. The landmark book is *The Scattering of Electromagnetic Waves from Rough Surfaces*, by Beckmann and Spizzichino.[4] Stover[5] has written a more recent book on optical scattering. Other books and articles on scattering theory and measurements are referenced in the anthology.[3]

1.2 What Surface Roughness Is

Surface roughness can take many forms. Often it consists of tiny scratches in random directions remaining after polishing, but it can also be the grooved structure produced by diamond-turning a metal mirror, the grain relief on a polished metal mirror such as molybdenum, the distinct random machining marks left on a finished metal part to give it a diffuse patina, the tiny parallel grooves on a glass surface that has been precision-ground, or even a few large scratches or pits (digs) sometimes caused by improper handling. Some materials such as ceramics or silicon carbide contain voids, while others—including aluminum or beryllium—have hard inclusions in their bulk; all these things can appear on the surface. Optical or magnetic films applied to surfaces can add additional roughness. With modern deposition techniques, optical dielectric films such as silicon dioxide, titanium dioxide, zinc sulfide, and magnesium fluoride are generally extremely smooth and contour the surfaces onto which they are deposited. However, silver, gold, and copper films, in particular, tend to be slightly lumpy, adding a fraction of a nanometer roughness to the surface. Other metal films such as aluminum, platinum, nickel, and rhodium are smoother since they are finer-grained. Magnetic films applied to aluminum or plastic surfaces and used to store information in dense arrays are generally quite rough compared with optical films. Optically black materials that absorb or scatter light are even rougher. These coatings have steep slopes and contain tiny holes acting as light traps because of multiple scattering.

The processing of a material can greatly affect its surface roughness. Here we give two examples, first a metal reflector and then a polished glass surface. For the metal reflector, consider a piece of polycrystalline copper such as oxygen-free, high-purity copper. This material can be cut to shape, rough finished, and then given a matte finish with a fine grade of sandpaper or emery paper. In this case, the surface looks slightly dull and contains a large number of scratches oriented in all directions. Alternately, the final finish can be made with a precision grinding wheel that

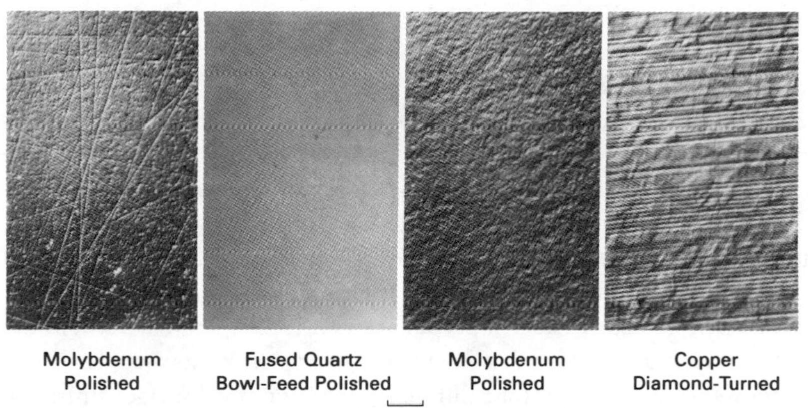

Figure 1.1 Nomarski micrographs of four different types of optical surfaces.

produces tiny parallel grooves. The surface looks burnished, but there are colored bands produced by diffraction from the grooves when the piece is viewed at certain angles. If the copper surface is optically polished on a pitch lap with a fine grade of abrasive in a liquid slurry, a smooth, shiny surface can be produced that contains tiny randomly oriented scratches and, sometimes, imbedded polishing compound. After polishing, the surface can be chemically etched to remove the subsurface damage generated by the polishing operation. Orange peel (waviness) is introduced by preferential etching of the grains and uneven etching of the entire surface. A grid of straight lines reflected from the surface appears wavy. Single-point diamond turning can produce a shiny surface with a minimum of subsurface damage. On the surface are closely spaced grooves made by the cutting diamond as well as coarser and deeper grooves caused by vibrations between the machine, tool, and surface; also, large grains in the material form a mottled surface structure (see Figure 1.1).

If aluminum, stainless steel, or beryllium had been considered, alloying material in the form of tiny hard particles dispersed throughout the material produce tiny bumps or pullouts (holes) in a polished surface. These increase the scattering level and, of course, the measured surface roughness.

In the second example, a piece of optical glass, which could be a window, beam splitter, filter, or witness sample, is first cut to shape by sawing or grinding. In some cases, glass can be molded to a size slightly larger than that of the finished shape and then fine-ground. The ground surface has a matte finish. Progressively finer grades of abrasive in water are used when the part is ground against an iron lap to bring the shape closer to the finished shape; the matte finish appears smoother. After the final grinding operation, polishing begins with a pitch lap and a fine abrasive mixed with water to make a slurry. Tiny surface asperities are polished flat but pits remain, giving the surface a gray appearance. As more material is removed, the pits gradually disappear, the grayness diminishes, and tiny scratches appear, oriented in all directions according to the random motion of the polishing machine.

To the unaided eye the surface appears shiny; the scratches can only be seen in a special differential interference contrast or Nomarski microscope (see Figure 1.1 and Section 1.4). To produce a superpolished surface, one can perform an additional polishing operation using a pitch lap and a slurry of very fine abrasive particles in water. The final polishing operation is done with pure water, and the only abrasive is what is imbedded in the lap. Polishing primarily occurs by a combination of mechanical and chemical actions. In the chemical part, water is the active reagent. During polishing, the outermost surface layer (perhaps a few atomic layers) dissolves and a thin film of pure silicon dioxide is redeposited, which covers up the tiny surface scratches. Thus, the final superpolished surface is scratch-free and extremely smooth. It appears featureless, even in the best optical microscope.

Figure 1.1 shows micrographs of four different types of optical surface finish, all taken with a differential interference contrast or Nomarski microscope (see Section 1.4). The featureless surface, polished fused quartz, is the smoothest with the lowest scatter. This type of surface finish is desired for the highest-quality optics. The three surfaces whose micrographs are shown on the right would all look shiny, while the scratched surface on the far left would have a grey cast caused by the scratches and pits.

The foregoing discussion applies to what surface roughness *is*. What it is *not* includes anything unintentionally added to a surface—fingerprints, dust, other particulates, surface contamination in the form of pollutants in the air, or oil films. Although each of these, in principle, can be removed from surfaces, it may be difficult or even impossible to remove them without damaging the underlying surface. Note that a surface can be intentionally covered with single or multiple

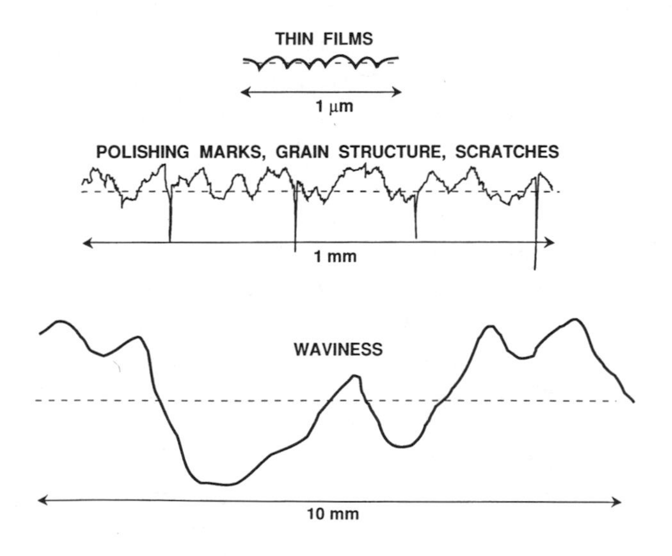

Figure 1.2 **Dimensions of various types of surface features. The height scale is ~1000 times smaller than the lateral scale.**

dielectric or metal films. These contour the underlying surface structure, adding tiny bumps in some cases, and may complicate the removal of unwanted surface contamination.

Surface roughness has two main attributes: roughness heights (or depths) and lateral dimensions. A scratch, for example, has a depth of a few tenths of a micrometer and a width of a few micrometers. The tiny crystallites that make up optical films have heights of a a few tens of nanometers and lateral dimensions of a fraction of a micrometer. Some types of film structure look like tiny pancakes on a surface. Figure 1.2 shows three types of surface features: thin films; polishing marks, grain structure, and scratches; and waviness or orange peel. The scale for the heights is orders of magnitude smaller than for the lateral dimensions. For example, for the middle drawing, the maximum peak-to-valley amplitude could be 100 nm and the maximum lateral dimension 1 mm = 1000 μm = 10^6 nm. The dimensions of surface features can range from atomic sizes to many micrometers.

Figure 1.3 shows the lateral dimensions of different types of surface features plotted on a logarithmic scale of surface spatial wavelengths which are separations of surface features or their lateral dimensions, as measured along the surface. Surface structure can be divided into three general groups according to the lateral dimensions. Surface microroughness (often called roughness) has lengths up to approximately 1 mm and includes thin films, polishing marks, scratches, and grain structure. Surface waviness, or mid spatial frequency roughness, has lengths from a few millimeters to perhaps 1 cm. Chemically polished surfaces such as those on silicon wafers exhibit mid spatial frequency roughness, commonly called orange peel. The overall surface shape, often called optical figure (departure from a perfect surface of the desired shape) or form in the machining industry, has lengths from

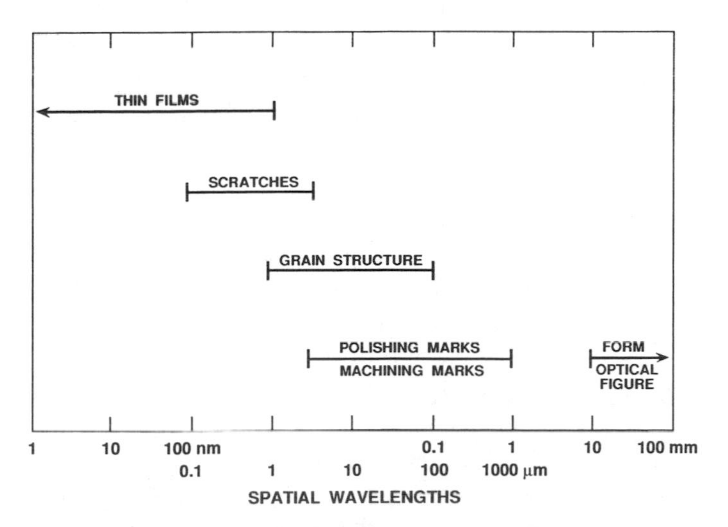

Figure 1.3 Graphs of the lateral dimensions of different types of surface roughness.

the centimeter range to the size of the piece, as indicated in Figure 1.3. Of course, some objects such as tiny lenses are smaller than 1 cm, so their figure and waviness will overlap. In this chapter, the emphasis is primarily on roughness rather than waviness or figure because we are mainly interested in the interaction between a rough surface and a light beam.

1.3 How Surface Roughness Affects Optical Measurements

All surface roughness scatters light. The character of the roughness—heights and spatial wavelengths—determines the intensity of the scattering and its angular distribution. As an example, an aluminum-coated glass mirror normally scatters ~0.1% of the light that it reflects. In other words, the ratio between the total scattered light and the total reflected light (specular reflectance plus scattering) is 0.1%. This surface would have a roughness of ~1.6 nm rms.

If we use this roughness as a ballpark figure, the total reflectance at normal incidence of an opaque mirror is reduced by about 0.1%. If light is reflected at nonnormal incidence, the specular reflectance is higher and the scattering losses lower. The transmission of a transparent piece of glass is also reduced by scattering from both surfaces.

If *intensities* are being measured, surface roughness scatters light out of the beam. However, if *polarization properties* are being measured, the effect of the surface roughness on a beam of plane polarized, circularly polarized, or elliptically polarized light at nonnormal incidence as in an ellipsometer or polarimeter[6] is negligible if the amount of surface roughness is small (i.e., the surface appears shiny).[7, 8] If the surface is so rough that multiple scattering occurs, the ellipsometric and polarimetric parameters are affected.[9]

Transmission measurements that depend on measuring an angle of deviation, as, for example, when determining the refractive index of a glass by measuring the angle of minimum deviation, are unaffected by surface roughness as long as it is small enough not to distort the slit image being observed. Striae, bubbles, and other types of inhomogeneities in the bulk of the glass are more serious problems in this type of measurement.

In waveguides where light is traveling within an optical fiber, surface roughness greatly affects the amount of light unintentionally coupled out of the fiber. In fact, this principle is used in the total internal reflection microscopy (TIRM)[10] technique to observe surface imperfections (see Section 1.5).

1.4 How Surface Roughness and Scattering Are Measured

A wide variety of methods are available for measuring surface roughness and the light scattering the roughness produces.[2] These methods can be grouped into those that give pictures of surfaces (and sometimes quantitative information) and those that yield quantitative statistical information about the surfaces. Figure 1.4 shows

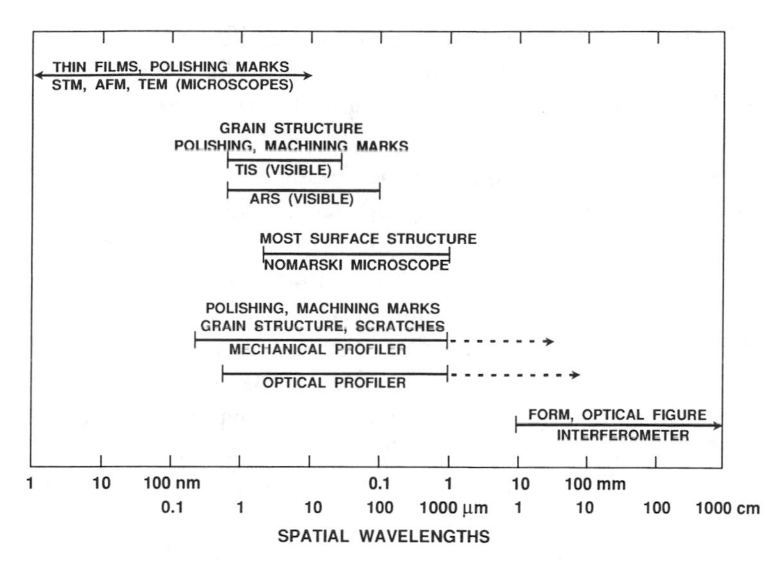

Figure 1.4 Techniques for measuring surface roughness in various spatial wavelength regions.

a diagram of measurement techniques that are suitable for various types of rough surfaces. This figure is similar to Figure 1.3 except that the measurement techniques have been added to the types of surface roughness. The various techniques are discussed in this section.

Among the methods that give pictures of surfaces are microscopes ranging from optical microscopes, scanning electron microscopes, and transmission electron microscopes to scanning probe microscopes. These latter instruments can be used to produce topographic maps of surfaces on an atomic scale, both laterally and vertically.[11, 12] If the surface is conducting, a scanning tunneling microscope (STM) can be used to produce a topographic map of the electron density. Most optical surfaces are nonconducting, and maps and profiles of the surface topography can be obtained by using an atomic force microscope (AFM). Some of the scanning-probe microscopes are now commercially available (from Digital Instruments, Inc., Park Scientific Instruments, and WYKO Corporation). The transmission electron microscope (TEM) and a newer scanning transmission electron microscope (STEM) are excellent for giving pictures of thin film surfaces or their cross sections. However, a surface replica must be made and shadowed at a suitable oblique angle in order to show the roughness structure on a film surface. Cross sectioning a film is even more difficult. The scanning electron microscope (SEM) requires steep surface slopes to produce an image with good contrast, and much effort and expertise are required to show the shallow but distinct structure on a diamond-turned metal surface.

A differential interference contrast or Nomarski (light) microscope[2, 13–15] is far superior to an SEM for observing roughness structure on smooth surfaces using magnifications ranging from 100× to 1000×. The lower magnifications are generally better because surface slopes are larger and the contrast in the image is better. Nomarski micrographs of four different types of optical surfaces are shown in Figure 1.1.

For smooth surfaces, two kinds of surface characterization methods give quantitative information about the surface roughness: optical and mechanical profilers that take profiles along a line or make topographic maps of an area, and light-scattering methods that can give information about surface statistical properties such as the rms roughness or power spectrum but no surface topography. Some of the general types of optical and mechanical profilers that are commercially available are listed in Table 1.1; they have been described in detail in Reference 2. Here we mainly comment on their similarities and differences. The obvious advantage of any optical profiler is that, since it is noncontact, it cannot damage the surface. All commercially available optical profilers are user-friendly, measurements can be taken easily and rapidly, and data reduction is rapid. The height sensitivities of the interferometer-based instruments are in the subnanometer range. The main disadvantages of the optical profilers are that (1) their lateral resolutions are limited by the properties of the optical systems and by the light beams illuminating the

Instrument	Type	Principle	Quantity Measured	Reference Surface
Bauer 100	Optical noncontact	Reflection of two light beams	Curvature	No
Chapman MP 2000	Optical noncontact	Differential interference contrast	Slope	No
Wyko TOPO-2D	Optical noncontact	Mirau interferometer	Phase of interference fringe	Yes
Zygo Maxim-3D	Optical noncontact	Fizeau or Mirau interferometer	Phase of interference fringe	Yes
Continental Optical Corp. Long Trace Profiler	Optical noncontact	Pencil-beam interferometer	Slope	No
UBM UB 16	Optical noncontact	Autofocus of CD optical head	Mechanical displacement	No
Rank Taylor Hobson Talystep	Mechanical contact	Diamond stylus	Mechanical displacement	No

Table 1.1 Commercially available optical and mechanical profilers.

surfaces and (2) the maximum step height that can be measured is less than half of the incident wavelength. The 1–2-μm lateral resolution is not sufficient to resolve fine scratches and other tiny surface details. As a consequence, roughness values obtained using optical profilers tend to be smaller than those measured with mechanical probe-type instruments on the same surfaces. Figure 1.5 illustrates this situation schematically. The wiggly line is a profile taken with a mechanical profiler using a 1-μm radius stylus, giving it submicrometer lateral resolution.[16] The bar graph and shading represent values that would be measured on the same surface using an optical noncontact profiler that had 2.5-μm lateral resolution. Note that the rms roughness value was decreased from 0.47 to 0.32 nm when going to the optical profiler. The height-measuring range, lateral resolution, and profile length of several commercially available profilers are shown in Figure 1.6.

Many other types of optical and mechanical profilers are described in the literature; about 30 different optical profilers are mentioned in References 2 and 3. Standards for calibrating optical and mechanical profilers are being developed at the National Institute for Standards and Technology (contact T. V. Vorburger) and at the National Physical Laboratory (contact A. Franks at the Division of Mechanical and Optical Metrology). These consist of surfaces having sine wave, square wave, or random profiles, or some combination of these, varying in amplitude and surface spatial wavelength.

Total integrated scattering (TIS) and angle-resolved scattering (ARS) can be used to make maps of scattering as a function of position on a surface and to give the

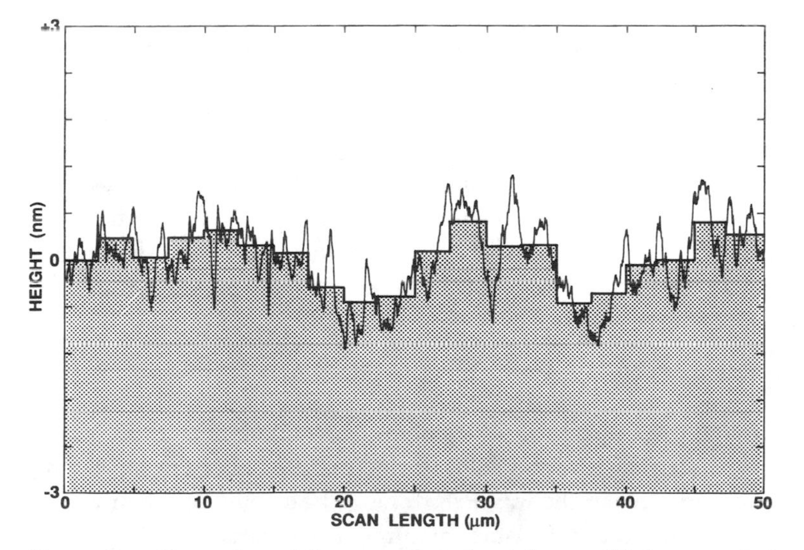

Figure 1.5 **Illustration of the averaging of a surface profile by an optical profiling instrument. The 0.47-nm rms original roughness was changed to 0.32-nm rms when the profile was averaged in 2.5-μm segments. (From Reference 2.)**

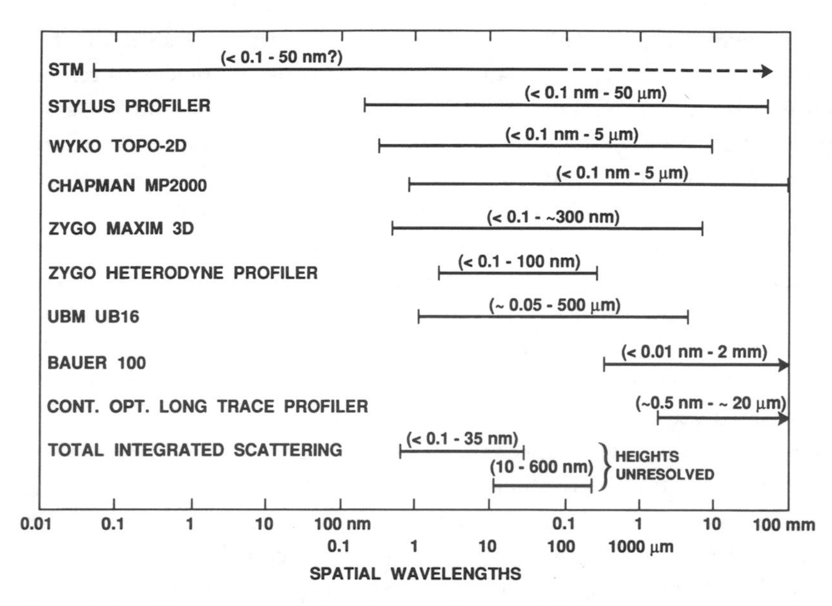

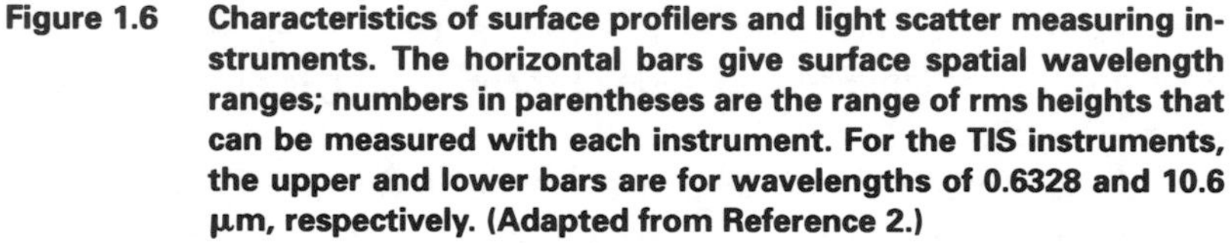

Figure 1.6 Characteristics of surface profilers and light scatter measuring instruments. The horizontal bars give surface spatial wavelength ranges; numbers in parentheses are the range of rms heights that can be measured with each instrument. For the TIS instruments, the upper and lower bars are for wavelengths of 0.6328 and 10.6 µm, respectively. (Adapted from Reference 2.)

rms roughness, power spectral density function, and other statistical properties of surfaces. The relation between scattering and the various statistical quantities depends on an assumption about the form of the surface roughness. If it consists of tiny polishing marks that are uniformly distributed over an isotropic surface, the roughness heights are small compared to the wavelength, and the lateral dimensions of the surface roughness are much larger than the wavelength, then scalar and vector scattering theories provide relations between the scattering and surface statistics.

The TIS from an opaque surface can be calculated from a simple relation obtained from scalar scattering theory. For light incident normally on the surface, the expression is[17, 18]

$$\mathrm{TIS} \approx (4\pi\delta/\lambda)^2 \tag{1.1}$$

where δ is the rms roughness and λ is the wavelength. This equation has been shown to predict correctly the scattering from an aluminized polished glass sample having a roughness of about 3 nm rms in the wavelength range from ~0.4 to 1.0 µm.[17] At present, there are no commercially available instruments for measuring TIS.

The expression for angle-resolved scattering derived from vector scattering theory is more complicated[19–21] and includes wavelength scaling (proportional to λ^{-4}), cosine of the angle of incidence, cosine squared of the scattering angle, functions of the optical constants of the surface and polarization of the incident and scattered light, and the surface power spectral density function. All quantities in the expression can be calculated from known parameters with the exception of the surface power spectral density function, which must be measured for the particular surface. Commercially available instruments (such as those from TMA Technologies) can be used to measure angle-resolved scattering at a variety of laser wavelengths in the visible and infrared spectral regions.

Many instruments for measuring TIS and ARS have been described,[2, 3] and there are over 100 articles reporting on TIS and ARS measurements in the open literature.[2, 3]

Unfortunately, the wavelength scaling of the scattered light for most surfaces does not obey either Equation 1.1 or the ARS relations at shorter wavelengths in the ultraviolet[22, 23] and longer wavelengths in the infrared. More theoretical and experimental studies are required in order for us to understand the dominant scattering mechanisms.

The TIS and ARS scatter measuring instruments are suitable for measuring surface features covering the entire surface such as microirregularities remaining from the polishing process. Isolated scratches and digs are not identified by these techniques. There is an official U.S. military specification MIL-0-13830A[24] for assessing the sizes of scratches by visually comparing them to "standard scratches" ruled on glass blanks. It has been shown by Young[25] and Johnson[26] that near-angle scattering from scratches of various widths can be duplicated by scattering from a group of closely spaced parallel grooves. This provides the possibility of having an objective reference with which to compare scattering from scratches on optical surfaces. There is also a commercial instrument (the image comparator microscope manufactured by Sira in England) that can be used to identify scratches by measuring the amount of light they scatter out of the collecting optics of the instrument.[27] Officially, though, scratches on optics in the U.S. must still be visually inspected according to the military specification.

Many machined surfaces have roughnesses whose heights are not small compared to the wavelength of the illuminating laser beam, generally a He–Ne laser. However, it is sometimes possible to use near-angle scattering to compare similar surfaces, for example, for quality control purposes, by calibrating the instrument with a surface whose roughness has been measured by a different instrument such as a mechanical profiler. A commercial instrument of this type is available.[28]

As mentioned previously, the interferometer-based optical profilers are limited to step heights of half of the incident wavelength. For rougher surfaces, there are optical stylus profilers (such as the UB 16 Precision Optical Length Measurement System manufactured by UBM-USA and the RM 600 Laser Stylus manufactured by Optische Werke G. Rodenstock) based on the auto-focusing system of the compact-disk optical head with a height sensitivity of ~0.05 μm, a lateral resolution of ~1 μm, and a dynamic range of up to 500 μm. These instruments provide a convenient extension of optical profilers into the rough surface domain of, for example, machined surfaces and paper surfaces, provided that the local slopes are less than a few degrees. A recent study[29] indicates that care must be taken in interpreting microroughness values measured on bulk scattering materials using autofocus instruments.

Confocal microscopy is another alternative for producing topographic maps of surfaces whose heights are large compared to the focal range (depth of focus) of a microscope objective, that is, $>\sim 1$ μm. A tiny area on a surface is imaged on a detector masked by a pinhole. If the area is in focus, the detector will have maximum signal output. If the area is out of focus, the detector signal will be lower. Two options are possible: (1) to move the imaging lens and detector until the signal is maximized and note the distance the lens was moved or (2) to calibrate the reduced signal level to correspond to the amount of defocus of the surface. Confocal optical microscopes have been used by the group at the University of Oxford for measuring surface topography whose heights are in the micrometer regime.[30–33] Other optical systems that act like pseudo-confocal microscopes and measure focus differences corresponding to surface height variations have been described by Sawatari and Zipin[34] and by Fainman et al.[35]

Surfaces that have steep-sloped roughness where multiple scattering can occur produce humps in the angle-resolved scattering curves at certain angles.[36–38] These same surfaces can enhance light scattered directly back into the incident beam by a factor of two, giving the so-called "opposition effect."[39] Theory predicts that enhancement in the retroscattered light can also occur when the surface roughness is much smaller and there are no multiple reflections.[40]

Although roughness values can be measured by a variety of techniques, care should be taken when comparing measurements made with different instruments since the surface spatial wavelength ranges may be different. Church has done extensive work in this area, starting in 1975, and has published nearly 50 papers on the subject. A bibliography of his papers is included in a recently published book of reprints.[3] Two of Church's recent papers with co-workers compare optical and mechanical measurements of surface finish[41] and discuss instrumental effects in surface finish measurement.[42]

1.5 Characterization of Selected Surfaces

Table 1.2 shows examples of six surfaces or groups of surfaces along with suggested ways of characterizing them. All of the materials in the first group (aluminum-coated glass mirror, diamond-turned copper mirror, and polished molybdenum mirror) are opaque and highly reflecting; they can all be characterized by using the same techniques, which have been discussed above. The surfaces can first be inspected in a Nomarski microscope to see whether they contain isolated defects that would affect the measurements or whether they require cleaning. If the diamond-turned copper mirror was turned on-center, it would have a small defect at that point that would affect profile or scattering measurements. The surfaces can be

Material	Characterization Technique
Aluminum-coated glass mirror Diamond-turned copper mirror Polished molybdenum mirror	Nomarski microscope for inspection or photograph Optical or mechanical profiler for surface profile Optical profiler and/or STM for topographic map TIS and ARS for scattering Interferometer for optical figure, shape, or form
Polished glass substrate Antireflection-coated window	Nomarski microscope for inspection Optical or mechanical profiler for surface profile Optical profiler and/or AFM for topographic map Interferometer for optical figure
Optical disk	Nomarski microscope (high magnification) for inspection or photograph Optical profiler (high magnification) or mechanical profiler (special stylus) for surface profile Scanning electron microscope? for photograph Optical profiler and/or AFM for topographic map
Magnetic disk	Nomarski microscope for inspection (surface cleanliness) or photograph Long-scan optical or mechanical profiler TIS for uniformity check
Optical black baffle	Nomarski microscope for inspection (surface uniformity) Infrared TIS or ARS for uniformity Reflectance versus wavelength for efficiency
Machined aluminum surface	Nomarski microscope for inspection (surface uniformity and defects) Optical or mechanical profiler for surface profile and surface statistics Infrared TIS or ARS for uniformity and statistics

Table 1.2 Suggested methods for characterizing different rough surfaces.

profiled with an optical noncontact profiler or a mechanical contact profiler to determine their rms roughness. If the aluminized glass mirror contains a network of tiny scratches remaining from the polishing process, the roughness value measured by the optical profiler will be smaller than that measured by the mechanical contact profiler; the latter can resolve tiny scratches with a sharp stylus (~1 μm radius and ~1 mg loading). One must be careful not to damage a soft surface (aluminum or copper) by using too large a loading. Molybdenum is harder, so scratching during mechanical profiling is not so much of a problem. Topographic maps can be made by using some types of optical profilers. If much higher lateral resolution is desired, the STM or AFM will give excellent topographic maps of small areas of the surface from fractions of a micrometer to ~100 μm on a side. The STM can only be used on conducting materials, so aluminum with its ~3 nm thick native oxide coating or slightly oxidized copper must be profiled with an AFM. TIS and ARS can give information about surface scattering; maps can be made of the scattering variations on the sample if it is possible to translate either the sample or the probe beam. If the surface contains only uniform, tiny scratches over the entire surface, the scattering measurements can be converted into an effective rms roughness (TIS, ARS), autocovariance function (ARS), and power spectral density function (ARS).[2] It is important that the theory being used to calculate surface roughness from scattering be valid for the type of roughness that is measured on the surface. Optical figure, shape, or form can be measured interferometrically using well-established techniques.

The second pair of surfaces in Table 1.2—glass substrates and antireflection coated windows—are transparent and have low reflectances. Thus, some techniques are inappropriate. A Nomarski microscope can show dirt, large scratches, and water marks, but is less sensitive to tiny scratches and surface microstructure since the reflectance is low. An alternate inspection technique, TIRM,[10] is much more sensitive but requires a laser, polarizer, coupling prism, and microscope. Profiles of the surfaces can be made with optical or mechanical profilers, and topographic maps can be made with some optical profilers. The AFM can also be used to map a surface on a much finer lateral scale. Optical figure can be measured interferometrically. TIS and ARS instruments are not useful since the samples are transparent and have low reflectance.

It is difficult to characterize glass or plastic lenses, particularly small ones or those that have small radii of curvature, because of their low reflectance, curved shape, and focusing properties. The surfaces can be visually inspected for obvious defects such as scratches and, if the curvature is not too strong, forward scattering can be measured either as TIS or ARS. However, since the scattering and reflections are coming from both surfaces and from the bulk of the material, it is not simple to determine an rms roughness value from the scattering measurement. A mechanical profiler is a good choice for obtaining surface roughness information because it is insensitive to the low reflectance of the surface.

Returning to Table 1.2, an optical disk contains tiny information pits, approximately ~1 μm wide and 0.2 μm deep. These can be seen in a Nomarski microscope at high magnification and can be profiled with an optical profiler at high magnification. A topographic map can also be made with one type of optical profiler. The pits can be profiled with a mechanical profiler using a special shovel-shaped stylus whose narrow dimension is ~0.1–0.3 μm. It may be possible to use an SEM to view the pits if the bare aluminum surface is studied before it is coated with transparent plastic. An AFM should give a good picture of the tiny pits.

The surface of a magnetic disk is rough by optical standards since the magnetic coating has steep-sloped grains. This type of surface can be inspected for uniformity in the Nomarski microscope, but neither an optical profiler nor a mechanical profiler has good enough lateral resolution to show the true shapes of the magnetic grains. However, because a long-scan optical profiler can be used to show the surface waviness and also any isolated large grains, it may be useful for quality control. TIS is good as a uniformity check if the entire surface can be mapped.

An optical black baffle is similar to a magnetic disk in that the black material is rough with steep-sloped structure. In addition to using a Nomarski microscope and infrared TIS or ARS to check for uniformity, one can measure the specular reflectance throughout the visible spectral region as well as in the infrared to be sure that it is low.

A machined aluminum surface is rough, but the surface slopes are much smaller than those for a magnetic disk or an optical black baffle. For this reason, it is possible to use an optical or mechanical profiler to get a good surface profile and surface statistics, and an optical profiler to obtain an area map. The Nomarski microscope is helpful for inspecting the surface to look for defects and check for cleanliness. TIS or ARS can be used with an infrared wavelength for obtaining surface statistics and checking for uniformity. The wavelength should be much larger than the roughness heights.

1.6 Future Directions

In the future, the STM, AFM, and other atomic probes will be used more to enable us to better understand surface roughness on an atomic scale and relate it to light scattering. There will be more long-scan profilers that have better lateral resolution and can measure surface heights in the waviness regime. Measurements made by these instruments will be used to improve the methods for making optical surfaces. Subsurface damage and anisotropic grain structure will be studied both from theoretical and from experimental points of view in order for us to better understand their effects on optical scattering and to minimize the scattering as much as possible. Finally, more theoretical work should be done for us to better understand the lack of wavelength scaling of the TIS and ARS into the ultraviolet and vacuum ultraviolet and into the infrared spectral regions.

References

1 *Rough Surfaces.* (T. R. Thomas, Ed.) Longman, London, 1982.

2 J. M. Bennett and L. Mattsson. *Introduction to Surface Roughness and Scattering.* Optical Society of America, Washington, DC, 1989.

3 J. M. Bennett. *Surface Finish and Its Measurement.* Optical Society of America, Washington, DC, 1992.

4 P. Beckmann and A. Spizzichino. *The Scattering of Electromagnetic Waves from Rough Surfaces.* Pergamon Press, London, 1963.

5 J. C. Stover. *Optical Scattering: Measurement and Analysis.* McGraw-Hill, New York, 1990.

6 R. M. A. Azzam and N. M. Bashara. *Ellipsometry and Polarized Light.* North-Holland, New York, 1977.

7 J. R. Blanco, P. J. McMarr, and K. Vedam. "Roughness Measurements by Spectroscopic Ellipsometry." *Appl. Opt.* **24**, 3773–3779, 1985.

8 J. R. Blanco and P. J. McMarr. "Roughness Measurements of Si and Al by Variable Angle Ellipsometry." *Appl. Opt.* **30**, 3210–3220, 1991.

9 I. Ohlídal and F. Lukes. "Ellipsometric Parameters of Rough Surfaces and of a System Substrate–Thin Film with Rough Boundaries." *Optica Acta.* **19**, 817–843, 1972.

10 P. A. Temple. "Total Internal Reflection Microscopy: A Surface Inspection Technique." *Appl. Opt.* **20**, 2656–2664, 1981.

11 H. K. Wickramasinghe. "Scanned-Probe Microscopes." *Sci. Am.* **261** (4), 98–105, 1989.

12 R. Pool. "The Children of the STM." *Science.* **247**, 634–636, 1990.

13 G. Nomarski. "Microinterférometre différentiel à ondes polarisées." *J. Phys. Rad.* **16**, 9S–13S, 1955.

14 G. Nomarski and A. R. Weil. "Application à la métallographie des méthodes interférentielles à deux ondes polarisées." *Rev. Metall.* (Paris). **52**, 121–134, 1955.

15 D. L. Lessor, J. S. Hartman, and R. L. Gordon. "Quantitative Surface Topography Determination by Nomarski Reflection Microscopy." *J. Opt. Soc. Am.* **69**, 357–366, 1979.

16 J. M. Bennett and J. H. Dancy. "Stylus Profiling Instrument for Measuring Statistical Properties of Smooth Optical Surfaces." *Appl. Opt.* **20**, 1785–1801, 1981.

17 H. E. Bennett. "Scattering Characteristics of Optical Materials." *Opt. Eng.* **17**, 480–488, 1978.

18 J. M. Bennett and L. Mattsson. *Introduction to Surface Roughness and Scattering.* Optical Society of America, Washington, DC, 1989, pp. 50–53.

19 J. M. Elson, H. E. Bennett, and J. M. Bennett. "Scattering from Optical Surfaces." In *Applied Optics and Optical Engineering,* Vol. 7. (R. R. Shannon and J. C. Wyant, Eds.) Academic Press, New York, 1979, pp. 191–244.

20 J. M. Elson and J. M. Bennett. "Vector Scattering Theory." *Opt. Eng.* **18**, 116–124, 1979.

21 J. M. Bennett and L. Mattsson. *Introduction to Surface Roughness and Scattering.* Optical Society of America, Washington, DC, 1989, pp. 53–55.

22 L. Mattsson, J. Ingers, and J. M. Bennett. "Wavelength Dependence of Angle-Resolved Scattering in the EUV-Visible. I. Experimental Results." *Appl. Opt.* 1992. In press.

23 J. Ingers, L. Mattsson, and J. M. Bennett. "Wavelength Dependence of Angle-Resolved Scattering in the EUV-Visible. II. Comparison Between Theory and Experiment." *Appl. Opt.* 1992. In press.

24 "Optical Components for Fire Control Instruments; General Specification Governing the Manufacture, Assembly, and Inspection Of." Military Specification MIL-0-13830A. 11 Sept. 1963. See especially the accompanying drawing No. 7641866, revision L, "Surface Quality Standards for Optical Elements (Scratch)."

25 M. Young. "Objective Measurement and Characterization of Scratch Standards." In *Scattering in Optical Materials.* (S. Musikant, Ed.) Proc. Soc. Photo-Opt. Instrum. Eng., Vol. 362, 1982, pp. 86–92.

26 E. G. Johnson, Jr. "Simulating the Scratch Standards for Optical Surfaces: Theory." *Appl. Opt.* **22**, 4056–4068, 1983.

27 L. R. Baker. "Inspection of Surface Flaws by Comparator Microscopy." *Appl. Opt.* **27**, 4620–4625, 1988.

28 R. Brodmann, O. Gerstorfer, and G. Thurn. "Optical Roughness Measuring Instrument for Fine-Machined Surfaces." *Opt. Eng.* **24**, 408–413, 1985.

29 L. Mattsson and P. Wagberg. "Assessment of Surface Finish on Bulk Scattering Materials. A Comparison Between Optical Laser Stylus and Mechanical Stylus Profilometers." *Precision Eng.* 1992. In press.

30 D. K. Hamilton and T. Wilson. "Surface Profile Measurement Using the Confocal Microscope." *J. Appl. Phys.* **53**, 5320–5322, 1982.

31 T. Wilson, A. R. Carlini, and D. K. Hamilton. "Images of Thick Step Objects in Confocal Scanning Microscopes by Axial Scanning." *Optik.* **73**, 123–126, 1986.

32 H. J. Matthews, D. K. Hamilton, and C. J. R. Sheppard. "Surface Profiling by Phase-Locked Interferometry." *Appl. Opt.* **25**, 2372–2374, 1986.

33 C. J. R. Sheppard and H. J. Matthews. "The Extended-Focus, Auto-Focus and Surface-Profiling Techniques of Confocal Microscopy." *J. Modern Optics.* **35**, 145–154, 1988.

34 T. Sawatari and R. B. Zipin. "Optical Profile Transducer." *Opt. Eng.* **18**, 222–225, 1979.

35 Y. Fainman, E. Lenz, and J. Shamir. "Optical Profilometer: A New Method for High Sensitivity and Wide Dynamic Range." *Appl. Opt.* **21**, 3200–3208, 1982.

36 K. A. O'Donnell and E. R. Mendez. "Experimental Study of Scattering from Characterized Random Surfaces." *J. Opt. Soc. Am. A.* **4**, 1194–1205, 1987.

37 E. R. Mendez and K. A. O'Donnell. "Scattering Experiments with Smoothly Varying Random Rough Surfaces and Their Interpretation." In *Scattering in Volumes and Surfaces.* (M. Nieto-Vesperinas and J. C. Dainty, Eds.) Elsevier, North-Holland, Amsterdam, 1990, pp. 125–141.

38 J. C. Dainty, N. C. Bruce, and A. J. Sant. "Measurements of Light Scattering by a Characterized Random Rough Surface." *Waves in Random Media.* **1**, S29–S39, 1991.

39 Z.-H. Gu, R. S. Dummer, A. A. Maradudin, and A. R. McGurn. "Experimental Study of the Opposition Effect in the Scattering of Light from a Randomly Rough Metal Surface." *Appl. Opt.* **28**, 537–543, 1989.

40 A. R. McGurn. "Enhanced Retroreflectance Effects in the Reflection of Light from Randomly Rough Surfaces." *Surf. Sci. Reports.* **10**, 357–410, 1990.

41 E. L. Church, J. C. Dainty, D. M. Gale, and P. Z. Takacs. "Comparison of Optical and Mechanical Measurements of Surface Finish." In *Advanced Optical Manufacturing and Testing II.* (V. J. Doherty, Ed.) Proc. Soc. Photo-Opt. Instrum. Eng., Vol. 1531, 1991, pp. 234–250.

42 E. L. Church and P. Z. Takacs. "Instrumental Effects in Surface Finish Measurement." In *Surface Measurement and Characterization.* (J. M. Bennett, Ed.) Proc. Soc. Photo-Opt. Instrum. Eng., Vol. 1009, 1988, pp. 46–55.

2

Characterization of the Near-Surface Region Using Polarization-Sensitive Optical Techniques

GERALD E. JELLISON, JR.

Contents

2.1 Introduction

Optical reflection techniques have long been recognized as premier characterization methods for the near-surface region of a material. Particularly, low-power optical measurements in the wavelength region from the near UV to the near IR (0.2–2.0 μm or 0.5–6.0 eV) offer several important advantages over competing diagnostic tools for surface and thin film analysis: optical probes are nondestructive, can be performed very rapidly, and can be carried out in any transparent ambient. Moreover, many polarization-sensitive measurements can be made very accurately and are very sensitive to changes in the dielectric function perpendicular to the surface. In certain situations, photons are the only possible probe. For example, ionized probes (such as electrons or He^{++} nuclei) are not feasible in plasma deposition systems or systems which employ magnetic fields, leaving only photons as a potential in situ monitor of the growth process. Another great advantage of near-visible optical measurements is that they need not be performed in a vacuum (such as experiments using electrons or He^{++} nuclei) and therefore the experimental system can be considerably simpler.

In spite of the many advantages, optical reflection characterization has some significant defects: (1) Optical probes are generally not sensitive to changes parallel to the surface of the sample; (2) Optical measurements often require extensive calibrations in order for the highest possible accuracy to be obtained; (3) The interpretation of optical data is dependent upon a model; and (4) The analysis depth is limited by the optical penetration depth of the probe light. These faults are generally not crippling, but do require that care be taken in the acquisition and analysis of the data.

In general, optical reflection measurements determine pseudodielectric functions of the sample being examined. If the system being measured consists simply of a semi-infinite homogeneous material and vacuum (i.e., with no interfacial region between the two media), then the measured pseudodielectric function is the dielectric function of the material. The deviation of the pseudodielectric function from the actual dielectric function of the material contains much information about the near-surface region. For example, single-wavelength nulling ellipsometry is the primary tool used in the semiconductor industry for determining the thickness and refractive index of films covering semiconductors. In addition, optical reflection techniques give information concerning surface roughness, film composition, interface layers, and crystalline quality.

Optical reflection measurements take two general forms. Reflectivity measurements are usually performed at near-normal incidence and determine only the intensity of the light reflected from the surface; no attempt is usually made to measure the polarization dependence of the reflected light or to measure the phase shift of the light upon reflection (although the phase is often obtained from reflectivity data from Kramers–Kronig analysis). It is difficult to get reflectivity measurements with accuracies much better than ~1%.

On the other hand, polarization-sensitive measurements directly determine some of the phase shift characteristics of the reflected light. The most important of these is ellipsometry, so named because the incident light onto the sample is elliptically polarized, and the object of the measurement is to determine the polarization ellipse of the light reflected from the sample surface. Ellipsometric measurements determine the ratio of the Fresnel reflection coefficients ρ, where

$$\rho = r_p / r_s = \tan\psi e^{i\Delta} \tag{2.1}$$

In Equation 2.1, $r_p(r_s)$ is the Fresnel reflection coefficient for light polarized parallel (perpendicular) to the plane of incidence. Recall that r_p and r_s are complex quantities; therefore, information concerning the phase shift of the light upon reflection is retained. Since the measured quantity is a ratio, the intensity of the incident light does not factor into the measurement and much greater accuracies can be obtained than with reflection intensity measurements.

Of course, several other optical techniques also are used to examine the near-surface region of materials. Optical microscopy images the surface of the material and is therefore good for measuring features in the plane of the surface. However, the resolution is limited by the wavelength of light, so feature sizes less that ~500 nm are generally not resolved. Optical absorption measurements determine the wavelength-dependent extinction coefficient of the entire material through which the light has traversed and is therefore generally not very sensitive to near-surface effects.

Optical reflection techniques are often sensitive to many of the same characteristics of the sample surface as cross section transmission electron microscopy (XTEM), Rutherford backscattering (RBS), and scanning tunneling microscopy (STM). XTEM is a more direct technique than optical reflection techniques, but it is destructive due to sample thinning. RBS measurements determine directly the depth of various atomic species and can give some information concerning crystalline quality; however, it is often difficult to distinguish between polycrystalline and amorphous material, and the depth resolution is limited to ~100 Å. STM, a relatively new technique, is very sensitive to features on the surface, but the technique requires a conducting sample, so measurements on low-conductivity materials such as amorphous silicon are difficult. All these techniques usually require the sample to be moved to the characterization experiment, eliminating the possibility for in situ measurements. Optical probes, on the other hand, can be used in situ for real-time monitoring of film growth, are not destructive, and can be used with samples of any conductivity, but the interpretation of the results is often model-dependent. Therefore, it is best to consider these techniques complementary; often, the best strategy is to examine at least some samples with more than one technique.

In this chapter, the focus is on ellipsometry techniques, which are a subset of reflection polarimetry. This technique has been the subject of considerable effort during the last 20 years, resulting in increased accuracy, ease of operation, and understanding in the interpretation of ellipsometry results. A brief discussion of each of the experimental techniques is followed by several examples of ellipsometry relating to the characterization of the near-surface region. In particular, comparisons are made with alternative diagnostic techniques, such as XTEM, when appropriate. The object of this chapter is to present, in abbreviated form, a wide variety of different applications of optical probes to the study of microstructure and the near-surface region of samples. No attempt is made to reference all the work done during the last 20 years; the reviews listed in the reference section contain a more complete list of references.

2.2 Ellipsometry

Experimental Implementations of Ellipsometry

In an ellipsometry measurement, elliptically polarized light is reflected from a sample surface at a large angle of incidence, and the polarization of the reflected beam is analyzed (see Figure 2.1 and Reference 1). An ellipsometer generally consists of a polarizer, a compensator, the sample, and an analyzing polarizer (the PCSA configuration), although some configurations place the compensator after the sample (PSCA). Over 100 years old, this is the standard technique in the semiconductor

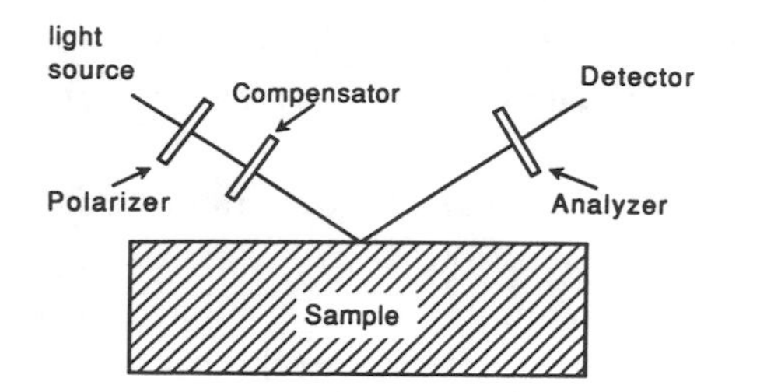

Figure 2.1 Schematic diagram of an ellipsometer.

industry for the measurement of the thickness of thin dielectric films on semiconductor surfaces.

The compensator or retarder plays an important role in the measurement capability of the ellipsometer. In general, this optical element changes linearly polarized light to elliptically polarized light (or vice versa), where the ellipticity of the emergent light depends upon the wavelength of light, the optical thickness of the compensator, and the azimuthal angle of the polarizer with respect to the fast axis of the compensator. Because the wavelength dependence of the compensator introduces complications, some implementations of ellipsometers do not use a compensator. However, the inclusion of the compensator in the instrument allows the instrument to measure quantities of the light reflected from a sample surface which otherwise cannot be determined.

The simplest form of an ellipsometric measurement is nulling ellipsometry. This measurement is generally performed at a single wavelength, and the compensating element is a quarter-wave plate. The azimuthal angle of the compensating element is fixed (usually at $\pm 45°$ with respect to the plane of incidence) and the azimuthal angles of the polarizer and analyzer are changed to minimize the light intensity at the detector. The ellipsometric angles ψ and Δ (see Equation 2.1) are determined directly from the readings of these azimuthal angles. Reference 1 gives an exhaustive description of this technique, including calibration, error analysis, and interpretation. Properly calibrated, the technique is very accurate, yielding errors in ψ and Δ of less than $0.01°$. However, the measurements are generally time-consuming, taking up to several minutes to perform one measurement.

During the last 20 years, ellipsometry instruments have been developed which use multiple wavelengths of source light. These spectroscopic ellipsometers generally come in two varieties: The first is the rotating analyzer (or polarizer) ellipsometer (RAE or RPE),[2–4] where the analyzer (or polarizer) is physically rotated, typically at less than ~100 Hz. The second is the polarization modulation ellipsometer (PME),[5–7] where the compensator is replaced with a photoelastic modulator (PEM), which modulates the ellipticity of the polarized light incident upon the sample, at a frequency ~50kHz.

In a typical RAE (or PRE) instrument, the light intensity incident upon the detector will be of the form

$$I(t) = I_0[1 + \alpha \cos 2(\omega t + \delta_0) + \beta \sin 2(\omega t + \delta_0)] \tag{2.2}$$

where $2\pi\omega$ is the frequency of the rotation of the optical element ($\lesssim$100 Hz). The Fourier coefficients α and β, along with the azimuthal angles of the fixed optical elements, are used to determine $\tan \psi$ and $\cos[\Delta + \delta_c(\lambda)]$, where δ_c is the wavelength-dependent retardation of the compensator. In most implementations of the RAE, the compensator is removed from the system, making $\delta_c(\lambda) = 0$. The lack of a compensating element in the RAE has both an advantage and a disadvantage: since the sample itself is the only wavelength-dependent optical element, the instrument is inherently simpler; however, without a compensator, the instrument can only measure quantities proportional to $\cos \Delta$. This type of RAE, then, is very insensitive to Δ for values close to 0° or 180° and cannot determine the sign of Δ. For many applications, the simplicity of the instrument more than justifies the limitation imposed by the elimination of the compensator. The RAE and RPE have been commercialized and are available from many companies today.

The other instrument mainly used for spectroscopic ellipsometry is the PME, which uses a PEM as the compensating optical element. The standard PEM consists of a drive crystal of piezoelectric crystalline quartz mechanically coupled to a piece of optical-quality fused silica that has a matched resonant frequency. An oscillating voltage across the faces of the drive crystal induces a periodic strain in the crystalline quartz, which is transferred to the optical fused silica. Due to the photoelastic effect, the sinusoidally varying strain in the fused silica results in a time-dependent refractive index, making it a time-dependent retardation device.

In all implementations of the PME, the light intensity reaching the detector will have a component that is independent of time and an oscillating component, which includes all the harmonics of the drive frequency of the PEM. The instrument measures the ratio of the fundamental and the second harmonic to the time-independent component, which are proportional to linear combinations of the associated ellipsometry parameters, given by

$$N = \cos 2\psi \tag{2.3a}$$

$$S = \sin 2\psi \sin \Delta \tag{2.3b}$$

$$C = \sin 2\psi \cos \Delta \tag{2.3c}$$

Although the inclusion of a PEM in the optical path complicates the instrument by adding a wavelength-dependent retardation, the wavelength dependence can be corrected electronically and it allows for the measurement of any of the three associated ellipsometry parameters. In one implementation,[7] all three can be measured simultaneously. The PME can measure both the magnitude and the sign of Δ very accurately, since both $\sin \Delta$ and $\cos \Delta$ can be measured. In addition, there is a built-in consistency check for the data, since $N^2 + S^2 + C^2 = 1$ (in some cases, $N^2 + S^2 + C^2 < 1$, indicating that the sample is depolarizing). Another advantage, which has not yet been practically exploited, is that the PME is about 500 times faster than the RAE. If it is assumed that the fastest possible measurement time is 5 times the time for a single cycle, then the fastest possible resolution time is ~100 μs for a PME, compared with ~50 ms for an RAE.

In all ellipsometry experiments, the probe light is incident upon the sample surface at a large angle of incidence ϕ. Since the Fresnel reflection ratio ρ (Equation 2.1) is extremely sensitive to ϕ, it is critical that this quantity be measured very accurately. In fact, one implementation of the spectroscopic ellipsometry, called variable angle of incidence spectroscopic ellipsometry (VASE),[8] utilizes this sensitivity to preselect the angle of incidence for the most sensitive value of ϕ (depending upon the sample being examined). Although this technique has been applied to RAEs, there is no reason why it could not also be applied to PMEs.

Recently, there has been considerable work on time-resolved ellipsometry experiments. Single-wavelength, in situ measurements have been made using an RAE[9] and a PME[10]; both of these systems have been used to follow the initial phases of film growth, and the results are described in more detail in the next section. In both systems, there is an inherent limitation of the time resolution due to the physical oscillation of the light polarization and to the integration time required for signal-to-noise ratio improvements; typical time resolutions are 0.5–3 s. Faster single-wavelength time-resolved ellipsometry can be obtained if an electrooptic modulator is used as the compensating element,[11] which results in a time resolution of ~4 μs. Even faster time resolution can be obtained if the sample is the only time-dependent polarizing/compensating element and the probe light source is a laser; the time resolution of this type of system[12–15] is limited by the resolution of the detector and data acquisition electronics and has demonstrated a time resolution of ~1 ns, with 1 ps resolution possible if the detection system is a streak camera.[16]

Recent work[17, 18] has shown that it is possible to combine an RAE with an optical multichannel analyzer, making it possible to do time-resolved, spectroscopic ellipsometry. Although the calibrations involved with this experiment are extensive and the time resolution is limited to ~3 s per spectrum, the potential for this instrument to aid in the understanding of the dynamics of slow film growth is enormous.

Analysis of Ellipsometry Data

The simplest analysis of ellipsometry data can be applied when the near-surface region of the sample can be approximated as a single interface, with a semi-infinite air or vacuum ambient and a semi-infinite substrate. In this case, the complex dielectric function of the substrate can be expressed as

$$\langle\varepsilon(\lambda)\rangle = \langle\varepsilon_1(\lambda)\rangle + i\langle\varepsilon_2(\lambda)\rangle = [\langle n(\lambda)\rangle + i\langle k(\lambda)\rangle]^2$$

$$= \sin^2\phi\left\{1 + \tan^2\phi\left[\frac{1-\rho(\lambda)}{1+\rho(\lambda)}\right]^2\right\} \tag{2.4}$$

where ϕ is the angle of incidence, ρ is the complex reflection ratio shown in Equation 2.1, and n and k are the refractive index and extinction coefficient, respectively. The angle brackets around ε, n, and k emphasize that Equation 2.4 can be used only to determine the pseudodielectric function; $<\varepsilon> = \varepsilon$ of the substrate only when the 2-medium approximation is valid.

For more complicated surface structures, a model of the sample surface must be constructed which will allow us to extract microstructural information concerning the near-surface region from the results of optical reflection experiments. If the sample has a relatively simple structure, then ψ and Δ can be used to determine two parameters in the sample surface model. For example, if the sample surface is a simple thin film of SiO_2 on silicon, then the results of a nulling ellipsometry experiment can be used to determine any two of the following parameters: oxide thickness, oxide refractive index, real or imaginary part of the substrate refractive index, or the angle of incidence, assuming that all the other parameters are known. This type of analysis works well only in restricted regions of ψ–Δ space, and it depends upon the choice of parameters to be determined (see Reference 1). There are several examples in the literature where spectroscopic ellipsometry data are interpreted in this manner, resulting in the determination of the wavelength dependence of the complex refractive index of the substrate.[19–24]

If the sample surface is more complicated, then a more sophisticated approach must be used to fit spectroscopic ellipsometry data. First, a model of the near-surface region must be assumed by approximating the surface as a series of layers. Second, the optical functions for each of the layers must be taken from the literature or parameterized using λ-independent quantities. Third, the experimental data $\rho_{exp}(\lambda)$ must be fit to calculated spectra $\rho_{calc}(\lambda)$ by varying film thicknesses or other parameters used to describe the optical functions of the individual layers.

In general, one minimizes the χ^2, where[25]

$$\chi^2 = \sum \frac{[\rho_{exp}(\lambda_i) - \rho_{calc}(\lambda_i, \mathbf{z})]^2}{\delta\rho^2(\lambda_i)} \tag{2.5}$$

The quantities $\rho_{exp}(\lambda_i)$ are the experimental complex ratios of the Fresnel reflection coefficients (see Equation 2.1) determined at i different wavelengths λ_i; the $\rho_{calc}(\lambda_i, \mathbf{z})$ are calculated assuming a specific model where the elements of the vector $\mathbf{z}$ are the quantities varied to minimize the χ^2. The components of the $\mathbf{z}$ vector can be film thicknesses, fractions of specific film constituents, or other parameters used to describe the complex refractive index of a film as a function of wavelength. The quantity $\delta\rho^2(\lambda_i)$ is the square of the experimental error in the measurement of $\rho_{exp}(\lambda_i)$. Until recently, most workers have assumed that $\delta\rho = 1$ and was independent of wavelength. This is clearly inappropriate for ellipsometry data taken using an RAE, since the accuracy of Δ becomes poor when Δ is close to 0° or 180°; many RAE workers have partially compensated by eliminating the data from consideration when $\cos\Delta \approx \pm 1$. The implications of ignoring the error limits have been discussed extensively in Reference 25.

There are several ways in which the optical functions of the individual layers can be parameterized. By far the most common is to assume that the layer consists of a microscopic mixture of two or more different phases and to determine the composite dielectric function an effective medium approximation.[26] In the Bruggeman effective medium approximation, the composite dielectric function ε is given by

$$\sum \frac{f_j(\varepsilon_j - \varepsilon)}{(\varepsilon_j + 2\varepsilon)} = 0 \tag{2.6}$$

where the sum j goes over all constituents of the film; f_j is the fraction of constituent j in the film, with dielectric function ε_j. Using this approximation, one can simulate surface features such as surface roughness and polycrystalline materials. Rough surfaces are usually approximated using Equation 2.6, assuming that one constituent is vacuum ($\varepsilon = 1$, to simulate voids) and the other constituent is the underlying material. Polycrystalline material is usually simulated using Equation 2.6, where one constituent is crystalline material and the other is amorphous material; in some cases, a third constituent, vacuum (or voids), is also included.

A second common parameterization is to assume that the optical functions of a layer can be represented as a sum of Lorentzian functions

$$\varepsilon - 1 = \sum \frac{A_j \lambda^2}{\lambda^2 - \lambda_j^2 + i\Gamma_j \lambda} \tag{2.7}$$

where the sum goes over all Lorentzian oscillators used to describe the optical functions of the material. Equation 2.7 is often used to simulate insulating films, where the linewidth $\Gamma_j = 0$; in this case, it is often called the Sellmeier approximation.

2.3 Microstructural Determinations from Ellipsometry Data

Temperature Dependence of the Optical Properties of Silicon

As a first example of the use of spectroscopic ellipsometry (SE) data, let us look at the determination of the optical properties of silicon at elevated temperatures. Although this feature is not strictly a microstructural characteristic, the temperature dependence of the optical functions of silicon influences many practical aspects of silicon technology. For example, these data were instrumental in understanding the proper mechanism of laser annealing.[27]

The temperature-dependent optical properties of silicon were measured by Jellison and Modine using spectroscopic PME.[19, 21, 22] In order to inhibit oxide growth of the silicon surface during the high temperature exposure, they bathed the sample continuously in a forming gas atmosphere (4% H, 96% Ar); by making room-temperature measurements before and after high temperature exposure, they ascertained that minimal additional oxide growth had occurred. Since the native oxide overlayer of Si is thin (~20 Å) and, for much of the wavelength region studied, the extinction coefficient of Si is small, the value of Δ can be very close to

180°; in this region, PME can measure Δ very accurately, so that the absorption coefficient $\alpha(= 4\pi k/\lambda)$ can be determined for $\alpha > 10^3$/cm. The temperature dependence of the absorption coefficient is shown in Figure 2.2, plotted semilogarithmically versus photon energy. Below the direct band gap (~3.4 eV), it has been empirically determined that

$$\alpha(\lambda, T) = \alpha_0(\lambda)\, e^{T/T_0} \tag{2.8}$$

where $\alpha_0(\lambda)$ is the value of the absorption coefficient at temperature T_0.

Determination of the Optical Functions of Glasses Using SE

The usual technique for the measurement of the refractive index of transparent glasses is the method of minimum deviation. The glass sample is cut in the shape of a prism, light enters the prism at a large angle of incidence, and the angle is found such that the angular deviation of the light exiting the prism is at a minimum. The refractive index is then found from this angle of deviation. With care, this technique can accurately determine refractive indices to the sixth place. There are, however, some disadvantages to this technique: it is extremely time-consuming, and

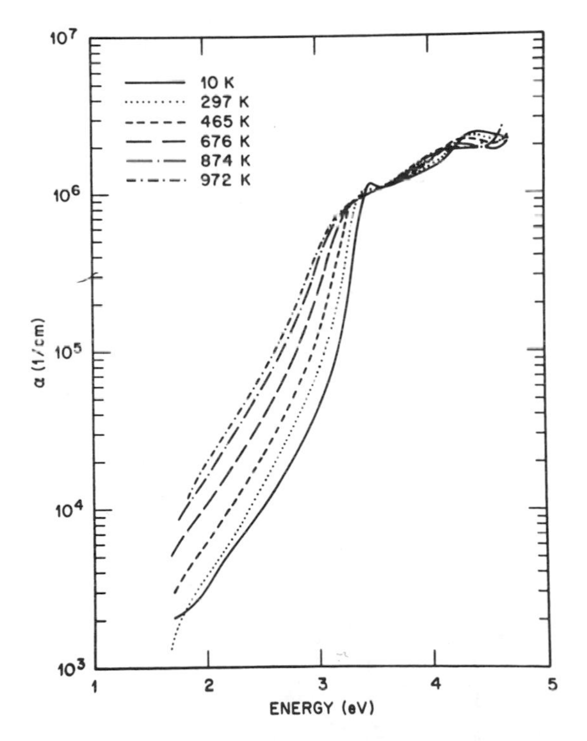

Figure 2.2 **The absorption coefficient of Si measured at several different temperatures using spectroscopic polarization modulation ellipsometry. (After Reference 21.)**

refractive index measurements can be made only at wavelengths where the material is transparent.

SE, using a PME instrument, offers an alternative method for the determination of the optical functions of glasses, even in regions where the material is opaque.[28] The sample must first be cut in the shape of a wedge; this deviates incident light that passes through the interface and is reflected from the back surface. Since Δ will be near 0° or 180° for transparent materials, accurate SE measurements can only be made with a PME instrument. The data is then analyzed using the 3-medium model (air/rough surface/glass), where the rough surface is approximated by Equation 2.6, using 50% air, 50% glass, and the refractive index and extinction coefficient of the glass are determined. The thickness of the interface layer is chosen such that the extinction coefficient is 0 at some long wavelength (such as 800 nm). The value of k determined at other wavelengths where the material is transparent serves as a self-consistency check, since the calculated value of k must be close to 0.

Figure 2.3 shows complex refractive index of a Pb–In–PO_4 glass, where the raw data has been converted to the pseudo-refractive index and pseudo-extinction coefficient (see Equation 2.1) and is labeled "Uncorr." Note that the uncorrected k is significantly greater than 0 from 1.5 eV to ~4.2 eV, indicating the presence of a rough layer on the glass surface. If the data are corrected for a 23 Å rough layer, then the corrected values of n and k are obtained and are shown in Figure 2.3. After the correction is made for the thin interface layer, the extinction coefficient reduces to 0 over the entire transparent region of the glass, indicating self-consistency. Note also that accurate values of the refractive index require that this thin interface layer must be properly taken into account, requiring that the measurements be taken with an instrument that can measure Δ accurately near 0° or 180°.

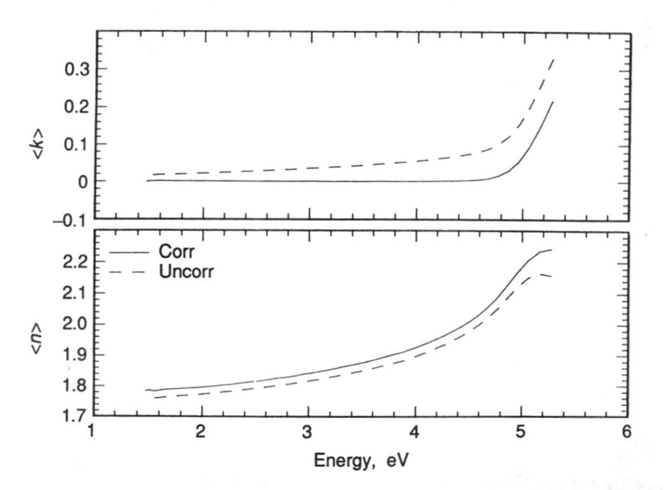

Figure 2.3 **The uncorrected (data) and corrected values of the refractive index *n* and the extinction coefficient *k* for an Pb–In–PO_4 glass. (After Reference 28.)**

Although the accuracy of the measurement of the refractive index (±0.002) does not compare with the accuracy of the minimum deviation method, this technique does present several distinct advantages: (1) the measurement is fast, allowing a complete spectroscopic measurement and data reduction in ~30 min, (2) the measurement is not limited to the transparent region of the spectrum, (3) only one surface must be polished, and (4) information is determined concerning the surface roughness layer, which would not otherwise be available.

Spectroscopic Ellipsometry Studies of SiO_2/Si

Generally, single-wavelength nulling ellipsometry experiments[1] measure the angles ψ and Δ directly and have been used for many years as a standard measurement technique for the thickness of a SiO_2 layer on Si. If it is assumed that the sample surface consists of a simple air/SiO_2/Si model, then the quantities ψ and Δ can be used to determine two of the following parameters: (1) film thickness, (2) film refractive index, (3) substrate refractive index, (4) substrate extinction coefficient, or (5) angle of incidence of the light onto the sample. This model assumes that there is no interface between the SiO_2 and the Si and that the extinction coefficient of the film is 0. For certain values of the oxide thickness and probe wavelength, two parameters may be mutually highly correlated, making it nearly impossible to calculate these two correlated parameters independently. For example, if the probe wavelength is 633 nm and the oxide thickness is very thin (less than ~10 nm), then the film thickness, film refractive index, and substrate extinction coefficient are all significantly correlated, as are the angle of incidence and the substrate refractive index; in this case, it is very difficult to determine both the film thickness and the film refractive index accurately from single-wavelength nulling ellipsometry measurements.

If the ellipsometry measurements are performed as a function of wavelength, then these correlations are minimized or eliminated; in addition, it is possible to examine more complicated models of the sample surface.[29, 30] In Reference 30, the near-surface region was modeled as air/SiO_2/interface/Si. Since it is not certain that the optical functions of the SiO_2 film grown on Si are the same as those of fused silica, the refractive index of the SiO_2 layer was modeled using the single-term Sellmeier approximation (Equation 2.7 with $\Gamma_1 = 0$). This formulation allows a very simple, yet realistic parameterization of the optical functions of film SiO_2; for fused silica, the values were found to be $A_1 = 1.099$ nm and $\lambda_1 = 92.27$ nm, resulting in an average absolute residual in n of 0.0002. The interface was assumed to be an effective medium, composed of a phase mixture of SiO_2 and Si. The values of the optical functions of Si were determined separately.[24] In this approximation, there are now five possible parameters: (1) the thickness of the SiO_2 layer, (2) the A_1 coefficient, (3) the λ_1 coefficient, (4) the thickness of the interface layer, and (5) the relative fraction of SiO_2 in the interface layer. The experimental spectra, including estimates of the experimental errors, are then fit to this model, where any combination of the five parameters can be varied; a good fit to the data is obtained when the biased $\chi^2 \approx 1$ (see Equation 2.5 and Reference 25). This fitting procedure also yields error limits for the parameters which are fit as well as cross-correlation coefficients that can be used to determine the degree of correlation between any of the two fitted parameters.

Table 2.1 gives the results of the fitting procedure for several SiO_2 films grown on Si using the normal dry oxidation procedure. Since the fraction of SiO_2 in the interfacial layer is strongly correlated with the interface thickness, this fraction is assumed to be 0.5. As can be seen from this table, very accurate values of the thickness are obtained. In addition, the Sellmeier coefficients A_1 and λ_1 are significantly different from the parameters associated with fused silica, indicating that the structure and refractive index of film SiO_2 is different from that of bulk SiO_2. For thicknesses below ~100 nm, the A_1 and λ_1 parameters are strongly correlated; therefore, it is not realistic to calculate both parameters and λ_1 is set to 92.27 nm. There also is a significant interface thickness for all samples, although thicker SiO_2 films tend to have thicker interfaces.

Spectroscopic Ellipsometry for Complicated Film Structures

SE has been used for years to study the near-surface region of many materials, some with very complicated layer structures.[31–34] In many cases, the information content is similar to that obtained by XTEM or RBS. SE is nondestructive and requires no special sample preparation, and the measurements can be taken and analyzed in a

$d(SiO_2)$, nm	A	λ_1, nm	d(interface), nm	χ^2
323.23 ± 0.23	1.1143 ± 0.0017	92.98 ± 0.39	0.92 ± 0.07	1.026
240.77 ± 0.20	1.1121 ± 0.0022	93.23 ± 0.44	0.55 ± 0.06	0.587
218.98 ± 0.19	1.1116 ± 0.0021	93.30 ± 0.44	0.61 ± 0.06	0.975
190.94 ± 0.10	1.1085 ± 0.0010	93.63 ± 0.46	0.64 ± 0.04	1.180
156.03 ± 0.08	1.1131 ± 0.0010	94.43 ± 0.53	0.76 ± 0.07	1.183
140.88 ± 0.10	1.1159 ± 0.0009	94.82 ± 0.46	0.75 ± 0.05	1.051
109.62 ± 0.05	1.1186 ± 0.0008	93.35 ± 0.57	0.52 ± 0.11	0.839
65.08 ± 0.03	1.1198 ± 0.0011	—	0.68 ± 0.05	0.836
42.37 ± 0.03	1.1264 ± 0.0041	—	0.32 ± 0.07	0.677
25.42 ± 0.05	1.1353 ± 0.0099	—	0.20 ± 0.08	1.012
Fused silica	1.099	92.27	—	—

Table 2.1 The values of the thickness, Sellmeier *A* and λ_1 coefficients for the SiO_2 layer, the thickness of the interface layer, and the resulting χ^2 of the fit. (After Reference 30.)

relatively short time (~1 h), so turnaround is significantly better than for XTEM. However, the interpretation of the data from SE experiments requires a realistic model to be assumed, from which parameters such as film thicknesses, interface roughness fractions, Sellmeier coefficients, etc., can be determined.

Figure 2.4 shows a direct comparison between SE and XTEM results from the same sample.[33] The sample was prepared by ion implantation of Si^+ ions into a crystalline Si substrate at 80 keV, a dose at 10^{16} ions/cm^2 and a beam current of 160 µA. As can be seen from the XTEM micrograph in Figure 2.4*a*, this sample preparation procedure yields a very complicated layer structure. Figure 2.4*b* shows the layer structure determined from XTEM; this technique is direct, but the sample examined is destroyed from the thinning process. Figure 2.4*c* shows the results obtained from a linear regression analysis of SE data from the same sample (before XTEM). The structure of the second and fourth layers are clearly complicated by the existence of multiple phases; the optical functions for these layers in the interpretation of the SE data were determined by an effective medium approximation (Equation 2.6), where the fraction of each of the constituents was allowed to vary and is shown in part *c* by the subscripts. Although SE is not a direct technique, it yields results which are in quantitative agreement with XTEM.

In a similar experiment, Woollam et al.[34] examined a thin film of $Al_xGa_{1-x}As$ grown on GaAs using molecular-beam epitaxy. Three experimental techniques were used: VASE, XTEM, and RBS. The thickness of the $Al_xGa_{1-x}As$ film was measured using all three techniques, the results agreeing within experimental error. A portion of this sample was then implanted with Ga ions and annealed. The VASE experiment indicated that the near-surface region was substantially single-crystal, while the RBS results indicated that part of the near-surface region was amorphous. XTEM results were able to clear up this apparent contradiction: the sample

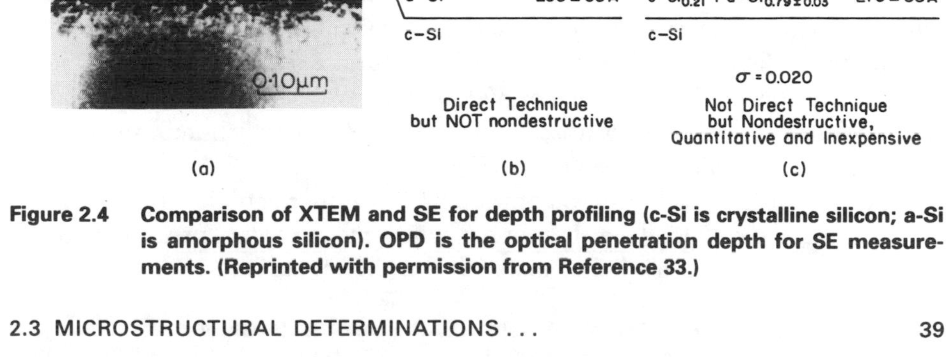

OPD C 0·10µm

(a)

XTEM		SE	
SiO_2	25 Å	SiO_2	24 ± 3 Å
c-Si + a-Si	120 ± 20 Å	c-$Si_{0.82}$ + a-$Si_{0.18\pm0.03}$	119 ± 19 Å
c-Si	550 ± 50 Å	c-$Si_{1.03\pm0.03}$	511 ± 21 Å
a-Si	250 ± 50 Å	c-$Si_{0.21}$ + a-$Si_{0.79\pm0.03}$	270 ± 30 Å
c-Si		c-Si	

$\sigma = 0.020$

Direct Technique but NOT nondestructive (b)

Not Direct Technique but Nondestructive, Quantitative and Inexpensive (c)

Figure 2.4 Comparison of XTEM and SE for depth profiling (c-Si is crystalline silicon; a-Si is amorphous silicon). OPD is the optical penetration depth for SE measurements. (Reprinted with permission from Reference 33.)

contained large-grain polycrystalline material which had many twin defects. Therefore, the SE (which is not sensitive to crystallographic direction for cubic materials) would see single-crystal material, while the incident He atoms in the RBS experiment would be randomly scattered from the misaligned crystallites due to the twinning. This experiment clearly illustrates the necessity of using many characterization techniques whenever possible.

Time-Resolved Ellipsometry

The fastest possible time-resolved experiments involve experimental systems where no optical elements move, either physically or electronically. Jellison and co-workers[12–15] have performed a series of time-resolved ellipsometry experiments during pulsed laser melting of silicon and germanium. These experiments were single-wavelength measurements, although different probe lasers were used to perform measurements at different wavelengths. The experimental setup was very similar to that shown in Figure 2.1, where a Babinet–Soleil compensator was used to circularly polarize the incident light; this device can quickly be adjusted to give circularly polarized light for any wavelength of light. A Wollaston prism was used for the analyzing prism, and both output beams were detected and digitized; thus, the data could be normalized, eliminating time variations of the probe beam intensity during the experiment. The highest time resolution obtained was ~1 ns, which was only limited by the speed of the photodetectors and the digitization rate of the electronics.

These experiments, along with several others[27] show high-energy laser pulses incident on a Si(Ge) surface result in silicon(germanium) surface melting. When the laser pulse first hits the sample, the surface heats. Once the surface temperature reaches the melting point (1410 °C for Si and 930 °C for Ge), a phase transition takes place and a pool of liquid material is formed on the surface. As more laser energy is absorbed into the near-surface region, the melt front rapidly goes into the sample. At some time, depending upon the characteristics of the laser pulse, the melt front returns to the surface, epitaxially regrowing single-crystal as it moves. The sample surface then cools to room temperature. Typically, the time required to melt the front surface is 10–20 ns, the melt duration is 20–300 ns, and the sample surface has cooled to room temperature in a few microseconds.

Time-resolved ellipsometric measurements of S (see Equation 2.3*b*) at a probe wavelength of 633 nm show that the S parameter never approaches 0 after the melt front has returned to the surface. This means that pulsed laser melting does not remove the surface oxide of Si, even though the surface temperature is at or above the melting point of Si for as long as 300 ns. However, if the Si sample surface is not precleaned, a one-time decrease in S can be observed, which indicates that the laser melting has removed organic contaminants.[12]

These measurements have also been used to measure the optical functions[13, 14] of liquid Si and Ge. Once the melt front has penetrated the optical penetration depth of the probe light into the liquid material, the ellipsometric response is constant

	Liquid Silicon		Liquid Germanium	
Wavelength, nm	*n*	*k*	*n*	*k*
633	3.80	5.20	3.26	5.92
514	3.11	4.89	2.51	5.12
488	2.94	4.99	2.39	4.86
458	2.74	4.96	2.08	4.73
364	1.93	4.32	1.69	4.01
351	1.80	4.26	1.56	3.91
334	1.74	4.09	1.46	3.76

Table 2.2 The values of the refractive index *n* and extinction coefficient *k* at selected wavelengths, determined by time-resolved ellipsometry. (After References 13 and 14.)

with time, until the melt front returns to the surface. During this time (usually 20–300 ns), the ellipsometric data can be simply interpreted as liquid material covered with a native oxide. Since the sample surface remains molten for such a short period of time, no additional oxidation can take place, making these measurements far more accurate than competing techniques. The values of the optical functions of liquid Si and Ge are shown in Table 2.2 for several probe wavelengths.

Time-resolved ellipsometry measurements, in conjunction with time-resolved reflectivity measurements, have shown that the melt-in phenomenon does not occur homogeneously.[15] That is, the melt front does not start at the front surface and move into the material as a planar melt front. Instead, the near-surface region (~20 nm) melts inhomogeneously, forming a mixture of liquid and solid Si. In cases where the energy density is just greater than the threshold energy density, the near-surface region never completely melts and still contains small microscopic regions of crystalline Si. For higher energy densities, the near-surface region starts out as an inhomogeneous melt until the fraction of crystalline Si goes to 0; then the melt front penetrates into the material as a planar melt front.

Single-Wavelength Real-Time Monitoring of Film Growth

Single-wavelength ellipsometry has recently been used to monitor the initial stages of film growth, with the emphasis on amorphous, microcrystalline Si, or polycrystalline Si on a silicon substrate.[9, 10, 35] Generally, a spectroscopic RAE or PME is attached to the growth chamber, allowing the user to select the probe wavelength. For Si, the optimum wavelength is ~3.4 eV (= 365 nm), where the light penetrates ~100 Å into c-Si. If the growth rate is slow, as is the case for many a-Si films, the 1–5 s time resolution of these experiments is sufficient to monitor the initial phases of film growth.

The data are usually plotted in $\langle\varepsilon_1\rangle$-$\langle\varepsilon_2\rangle$ space, which will follow a specific trajectory, depending upon the growth kinetics. Figure 2.5 shows a typical growth trajectory for the glow-discharge deposition of amorphous Si from (*a*) pure SiH_4 with a substrate temperature T_s of 250 °C and (*b*) from a 1:10 flow ratio of SiH_4:H_2 with a T_s of 240 °C. The solid line in Figure 2.5*a* is based on a model assuming convergence of initial growth microstructure after 50 Å, leaving 6 Å of roughness at the film surface, increasing to 8 Å in the later growth stages. In Figure 2.5*b*, the solid line is calculated assuming convergence of initial growth microstructure after ~37 Å, leaving behind a 50-Å interface layer of lower Si–Si bond-packing density. The beginning point (upper right corner of each figure) corresponds to the pseudodielectric function for the substrate (oxidized, crystalline Si), while the convergence point corresponds to the pseudodielectric function for the film (oxidized, amorphous Si). If the growth were planar, monolayer by monolayer, then the dotted trajectory shown in the figure would be followed. The existence of the cusp indicates that the growth is not uniform, but rather favors island formation, which coalesces into a uniform film by the time the data trajectory reaches the cusp. At later times, features such as the density of amorphous Si and the surface roughness determine the particular path that the experimental data will take in $\langle\varepsilon_1\rangle$-$\langle\varepsilon_2\rangle$ space.

This work, along with many other examples in the literature, show the enormous potential for in situ monitoring of film growth. The trajectories produced in these studies are sensitive to surface roughness, interface roughness, as well as the crystal quality of the film itself. Since the probe itself is nonintrusive, this technique can be used to monitor film growth from virtually any process, including plasma deposition or magnetron sputtering.

Recent work by Aspnes et al.[36, 37] has shown that real-time in situ monitoring of the ellipsometric response of a surface can be used to control the parameters for film growth. This use of ellipsometry is clearly important from a technological point of view, in that it promises greater control of surface quality and film thickness.

In the experiments of Aspnes et al., a rotating polarizer ellipsometer was attached to a organometallic molecular beam epitaxy system for the growth of $Al_xGa_{1-x}As$ thin films on a GaAs substrate. The ellipsometer was operated at 2.6 eV (475 nm) because the imaginary part of the dielectric function ε_2 varies very strongly with x at this wavelength. A closed-loop system was constructed to keep ε_2 of the $Al_xGa_{1-x}As$ film to a precision of better than ±0.001 over an extended period of time, where the flow of triethylaluminum was regulated from the optically measured value of ε_2.

Multiple-Wavelength Real-Time Monitoring of Film Growth

Although the time-resolved single-wavelength ellipsometry experiments discussed in the previous section provide invaluable information concerning the growth of thin films, the results require that a model of the sample surface as a function of time be assumed; the time-evolution of this predicted model is then compared with the observed trajectory in $\langle\varepsilon\rangle$ space. There is no consistency check with the data,

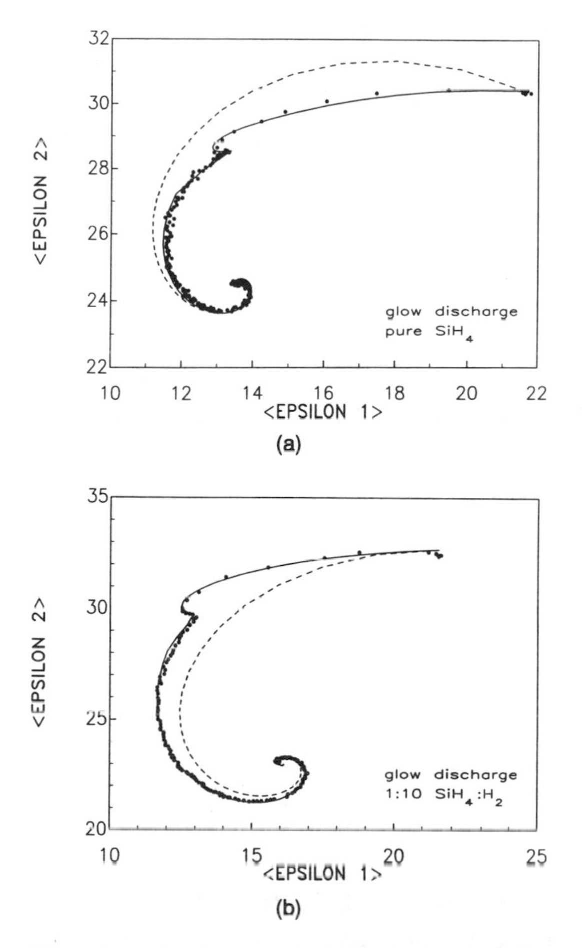

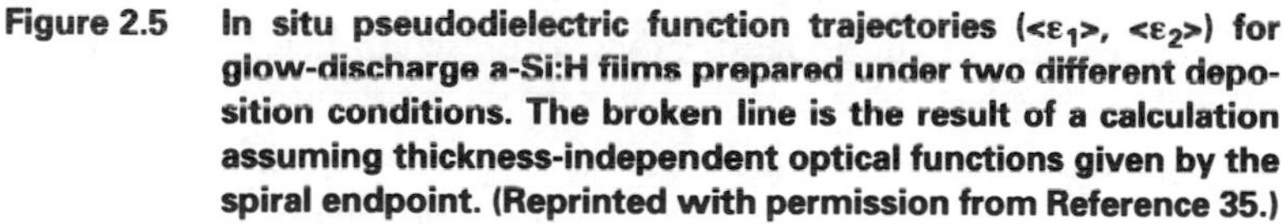

Figure 2.5 **In situ pseudodielectric function trajectories ($<\varepsilon_1>$, $<\varepsilon_2>$) for glow-discharge a-Si:H films prepared under two different deposition conditions. The broken line is the result of a calculation assuming thickness-independent optical functions given by the spiral endpoint. (Reprinted with permission from Reference 35.)**

and it may be possible that many other models work as well. To get more information concerning the growth process, Collins and co-workers[17, 18, 38] have developed real-time spectroscopic ellipsometry using the rotating polarizer configuration, where the exit light is detected spectroscopically using an optical multichannel analyzer. These experiments determine the complete pseudodielectric function from 1.5 to 4.5 eV with a resolution of 3 s and a repetition period of 15 s. This system is ideal for the study of the slow growth of amorphous Si, which grows ~0.1 Å in the 3 s resolution time in their experiments.

Figure 2.6 shows the real and imaginary part of the pseudodielectric function for amorphous Si growing on a crystalline substrate. At the beginning of growth,

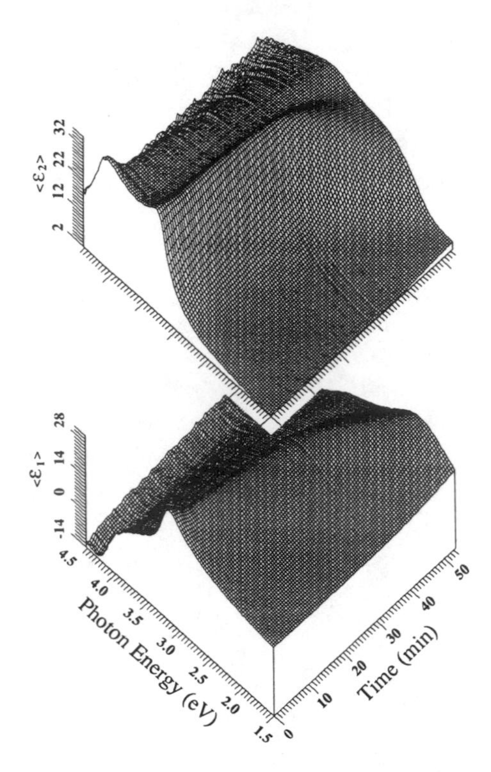

Figure 2.6 **Psuedodielectric function obtained during the growth of amorphous silicon on a crystalline silicon substrate. (Adapted with permission from Reference 38.)**

the pseudodielectric function is just that of Si with an overlayer of SiO_2 of ~20–25 Å. As the growth is begun, the pseudodielectric function is slowly modified by the amorphous Si film until, at the longest times, the pseudodielectric function resembles that of bulk amorphous Si.

The analysis of this data is extremely time-consuming, since a complete regression analysis must be performed on each spectrum; thus, the analysis can only be performed after the experiment has been concluded. For the data shown in Figure 2.6, this detailed fit indicated that the growth proceeded in two steps: first, nuclei were formed, until the film was ~13 Å thick; at this point, there was an abrupt transition to bulk film growth.

Infrared Ellipsometry Studies of Film Growth

All the techniques discussed so far have used photons in the near-UV to near-IR range, which is not sensitive to the vibrational characteristics of the atomic species in the near-surface region. Recent work by Drévillon and co-workers[39–41] describes a new infrared ellipsometer based on the PME design. Since this is a reflection technique, many of the drawbacks of infrared transmission spectroscopy are avoided, such as the need to have a transparent substrate (in the IR) and the difficulties involved in the interpretation of the results of thin films on a substrate. In addition, this technique has all the advantages of normal ellipsometry measurements: it is very sensitive, it is a ratioing technique avoiding problems with a fluctuating source intensity, and it measures two quantities (such as ψ and Δ).

Reference 40 describes IR ellipsometry measurements made on a series of ultrathin hydrogenated amorphous silicon films. The IR ellipsometric technique was able to detect the presence of Si–H and Si–H_2 bond-stretching modes in the thickest films examined (500 Å); however, these stretching mode vibrations were not observed for the very thinnest films. These results indicate that there is a very strong chemical interaction between the substrate and the first few monolayers of the amorphous silicon film, resulting in the interface region being significantly different from the bulk of the film. In another set of experiments,[41] IR ellipsometry spectra were taken in situ during growth of amorphous silicon on fused silica and B_2O_3-containing glass substrates. The experiments on films grown on the glass substrate showed evidence for atomic vibrations of BO groups which were not in evidence for films of amorphous silicon grown on fused silica. Clearly, boron has been chemically incorporated into the interface of amorphous Si films grown on glass substrates.

Acknowledgments

The author would like to thank F. A. Modine and D. H. Lowndes for carefully reviewing this manuscript. The writing of this review was sponsored by the Division of Materials Sciences, U.S. Department of Energy under contract DE-AC05-84OR21400 with Martin Marietta Energy Systems, Inc.

References

1 R. M. A. Azzam and N. M. Bashara. *Ellipsometry and Polarized Light.* North-Holland, Amsterdam, 1977.

2 D. E. Aspnes and A. A. Studna. *Appl. Opt.* **14**, 220–228, 1975.

3 R. W. Collins. *Rev. Sci. Instrum.* **61**, 2029–2062, 1990.

4 G. H. Bu-Abbud, N. M. Bashara, and J. A. Woollam. *Thin Solid Films.* **137**, 27–41, 1986.

5 S. N. Jasperson and S. E. Schnatterly. *Rev. Sci. Instrum.* **40**, 761–767, 1969.

6 B. Drevillon, J. Perrin, R. Marbot, A. Violet, and J. L. Dalby. *Rev. Sci. Instrum.* **53**, 969–977, 1982.

7 G. E. Jellison, Jr. and F. A. Modine. *Appl. Opt.* **29**, 959–974, 1990.

8 G. H. Bu-Abbud, N. M. Bashara, and J. A. Woollam. *Thin Solid Films.* **137**, 27–41, 1986.

9 R. W. Collins. "Ellipsometric Study of α-Si:H Nucleation, Growth, and Interfaces." In *Advances in Amorphous Semiconductors.* (H. Fritzsche, Ed.) World Scientific, Singapore, 1989, p. 1003.

10 A. Canillas, E. Bertran, J. L. Andújar, and B. Drévillon. *J. Appl. Phys.* **68**, 2752–2759, 1990.

11 A. Moritani, Y. Okuda, H. Kubo, and J. Nakai. *Appl. Opt.* **22**, 2429, 1983.

12 G. E. Jellison, Jr., and D. H. Lowndes. *Appl. Opt.* **24**, 2948–2955, 1985.

13 G. E. Jellison, Jr., and D. H. Lowndes. *Appl. Phys. Lett.* **47**, 718–721, 1985.

14 G. E. Jellison, Jr., and D. H. Lowndes. *Appl. Phys. Lett.* **51**, 352–354, 1987.

15 G. E. Jellison, Jr., D. H. Lowndes, D. N. Mashburn, and R. F. Wood. *Phys. Rev. B.* **34**, 2407–2415, 1986.

16 G. E. Jellison, Jr. *Opt. Lett.* **12**, 766–768, 1987.

17 Y. T. Kim, D. L. Allara, R. W. Collins, and K. Vedam. *Thin Solid Films.* **193**, 350–360, 1990.

18 N. V. Nguyen, B. S. Pudliner, I. An, and R. W. Collins. *J. Opt. Soc. Am. A.* **8**, 919–931, 1991.

19 G. E. Jellison, Jr. and F. A. Modine. *J. Appl. Phys.* **53**, 3745–3753, 1982.

20 D. E. Aspnes and A. Studna. *Phys. Rev. B.* **27**, 985–1009, 1983.

21 G. E. Jellison, Jr. and F. A. Modine. *Appl. Phys. Lett.* **41**, 180–182, 1982.

22 G. E. Jellison, Jr. and F. A. Modine. *Phys. Rev. B.* **27**, 7466–7472, 1983.

23 P. Lautenschlager, M. Garringa, L. Viña, and M. Cardona. *Phys. Rev. B.* **36**, 4821–4830, 1987.

24 G. E. Jellison, Jr. *J. Opt. Mat.* **1**, 41–47, 1991.

25 G. E. Jellison, Jr. *Appl. Opt.* **30**, 3354–3360, 1991.

26 D. A. G. Bruggeman. *Ann. Phys. Leipzig.* **24**, 636, 1935.

27 *Pulsed Laser Processing of Semiconductors.* (R. F. Wood, C. W. White, and R. T. Young, Eds.) Semiconductors and Semimetals, Vol. 23. 1984.

28 G. E. Jellison, Jr., and B. C. Sales. *Appl. Opt.* **30**, 4310–4315, 1991.

29 D. E. Aspnes and J. B. Theeten. *J. Electrochem. Soc.* **127**, 1359–1365, 1980.

30 G. E. Jellison, Jr. *J. Appl. Phys.* **69**, 7627–7634, 1991.

31 D. E. Aspnes. *SPIE Proceedings.* **276**, 188–195, 1981.

32 D. E. Aspnes. *SPIE Procedings.* **946**, 84–97, 1988,

33 K. Vedam, P. J. McMarr, and J. Narayan. *Appl. Phys. Lett.* **47**, 339–341, 1985.

34 J. A. Woollam, P. G. Snyder, A. W. McCormick, A. K. Rai, D. Ingram, and P. P. Pronko. *J. Appl. Phys.* **62**, 4867–4871, 1987.

35 R. W. Collins and J. M. Cavese. *J. Appl. Phys.* **62**, 4146–4153, 1987.

36 D. E. Aspnes, W. E. Quinn, and S. Gregory. *Appl. Phys. Lett.* **56**, 2569–2571, 1990.

37 D. E. Aspnes, W. E. Quinn, and S. Gregory. *Appl. Phys. Lett.* **57**, 2707–2709, 1990.

38 I. An, H. V. Nguyen, and R. W. Collins. *Phys. Rev. Lett.* **65**, 2274–2277, 1990.

39 R. Beneferhat and B. Drévillon. *Thin Solid Films.* **156**, 295–305, 1988.

40 B. Drévillon and R. Benferhat. *J. Appl. Phys.* **63**, 5088–5091, 1988.

41 N. Blayo and B. Drévillon. *Appl. Phys. Lett.* **57**, 786–788, 1990.

3

The Composition, Stoichiometry, and Related Microstructure of Optical Materials

ROBERT J. NEMANICH and TREVOR P. HUMPHREYS

Contents

3.1 Introduction

An important aspect of opto-electronic engineering is the alloy concentration dependence of the bandgap of semiconducting alloys. By the control of alloy concentration, the appropriate optical response can be engineered for complex thin film structures. Another aspect that can contribute to the optical response involves deviations from stoichiometry of the compound. In general, stoichiometry deviations are related to defects or impurities. For both alloy concentration and nonstoichiometry, the atomic bonding and microstructure will determine the specifics of the effects.

Although significant advances in the chemical analysis of optical materials have recently been reported,[1] this review addresses methods of characterizing the alloy and stoichiometry dependence of the optical response of thin films. Although there are many applications, we focus on three groups which represent important and typical situations. These include III–V materials which are often used for semiconducting light emitters or detectors, group IV epitaxial semiconductors with applications as integrated light detectors, and amorphous and microcrystalline materials with applications ranging from xerography to photodetectors.

Measuring alloy concentration or stoichiometry or both is often difficult. Although there have been significant advances in measuring the alloy composition and stoichiometry of bulk samples, most of these techniques cannot be applied to thin films. For thin films, surface probes such as Auger spectroscopy (AES), X-ray photoemission spectroscopy (XPS), secondary ion mass spectroscopy (SIMS), and Rutherford backscattering spectroscopy (RBS) have sensitivity to the elements but difficulty in obtaining relative concentrations to better than 10%. Small deviations of stoichiometry can also lead to significant changes in the optical properties, particularly in the optical absorption for subband gap light due to defect and impurity levels. Although these measurements themselves can sometimes be used to characterize the deviations of stoichiometry, it is usually difficult to be specific as to the particular atomic arrangements which lead to the effects.

A method often used to characterize optical materials involves measuring aspects related to the chemical bonding structure. In general, the optical bandgap of a material will be directly related to the local atomic bonding. Raman scattering has proven to be a particularly useful technique for analysis of the local bonding structure. Some of the most successful studies of the effects of alloy concentration and stoichiometry have correlated the optical absorption and Raman scattering to atomic structure measurements.

The emphasis of this text is on reviewing the applications of Raman scattering for the characterization of the alloy concentration and stoichiometry of some optical materials. The particular focus is on a group of materials often considered for opto-electronic applications ranging from integrated opto-electronic structures, semiconducting light emitters, xerography, and solar cells. A comprehensive review of all possible applications and results even within this selected group of topics is impossible within the scope of this chapter.

3.2 Aspects of Raman Scattering

The measurements and analyses described in this chapter are based on the laser Raman scattering technique.[2, 3] The technique involves inelastic scattering of light using a laser tuned to a single frequency. The scattered light is spectroscopically analyzed using a grating monochromator and sensitive photodetector. In all the measurements described here, we are concerned only with Raman scattering from the vibrational modes of the materials. Effects involving other types of excitations, including electronic excitations, are not considered. In addition, most of the experiments are carried out under nonresonant conditions. This means that the energy of the incident or scattered photons will not correspond to distinct electronic transitions in the material. Although resonant scattering can enhance the sensitivity of the technique, it often leads to strong variations of the Raman scattering efficiency of different components in the spectrum. Since the general goal of the measurements described here is to correlate the features of the spectrum to composition aspects of the material, nonresonant conditions are usually preferred.

In general, the Raman scattering experiments involve measurements of semiconducting thin films which are not transparent to the incident radiation. Thus, a backscattering geometry is often employed. To minimize sample heating, one may focus the incident laser light with a simple cylindrical lens to an image of ~2 mm × 100 μm. Since most of the measurements involve oriented crystalline structures, polarization selection rules can and must be considered. This means that all measurements must recognize the potential for variations depending on sample orientation with respect to the polarization configuration of the incident and scattered light.

3.3 III–V Semiconductor Systems

To date, first-order Raman scattering of III–V compounds, in particular GaAs and the corresponding ternary alloy AlGaAs, has been widely reported.[4, 5] Since Raman scattering is a nondestructive local probe of the lattice dynamics of the crystal, important information pertaining to the local crystalline structure, composition, mechanical strain, impurity related defects, alloying, and doping levels is provided. Moreover, since the penetration depth of the laser light into the semiconductor material can be easily varied, the technique is a versatile tool to investigate thin epitaxial layers grown on dissimilar substrates, for example, GaAs on Si. In addition, recent studies of the thin film growth of III–V nitrides (GaN, AlN, and BN) have routinely employed Raman scattering measurements to determine the crystallographic phase of the deposited layer.[6–8]

In general, the first-order spectrum from an undoped, semi-insulating single crystal of GaAs or InP shows two peaks corresponding to transverse-optical (TO) and longitudinal-optical (LO) phonon scattering.[9–11] For GaAs, the respective phonon frequencies are 269 and 292 cm^{-1} with corresponding linewidths of less than 3 cm^{-1}. In the backscattering geometry on the (100) surface, the TO phonon mode is forbidden according to polarization selection rules. However, its presence has been widely reported for both GaAs (100) and InP (100) single crystals and corresponding heteroepitaxial films.[12–14] Several interpretations regarding its origin, including the experimental deviation from the strict backscattering geometry and the breakdown of symmetry selection rules due to electron scattering by ionized dopant impurities, have been proposed.[15] In contrast, it has been suggested by Fontaine et al.[16] and Landa et al.[17] in a study of heteroepitaxial GaAs (100) films grown on (Ca,Sr)F_2 substrates that the presence of the TO mode is the result of disordered-induced scattering in the near surface region due to microstructural defects in the GaAs layer. In particular, as evidenced by TEM analysis, the TO phonon mode is related specifically to the presence of microtwins in the GaAs epitaxial layer. Indeed, it has been proposed that a quantitative measurement of the volume of epitaxial film that is twinned is related to the intensity ratio of the first-order optical phonon modes (I_{TO}/I_{LO}). Therefore, the ratio of the TO and LO amplitudes in the Raman spectrum gives a quantitative measure of crystal quality. Similar results have been reported by Humphreys et al.[18] for the heteroepitaxial growth of GaAs (100) films

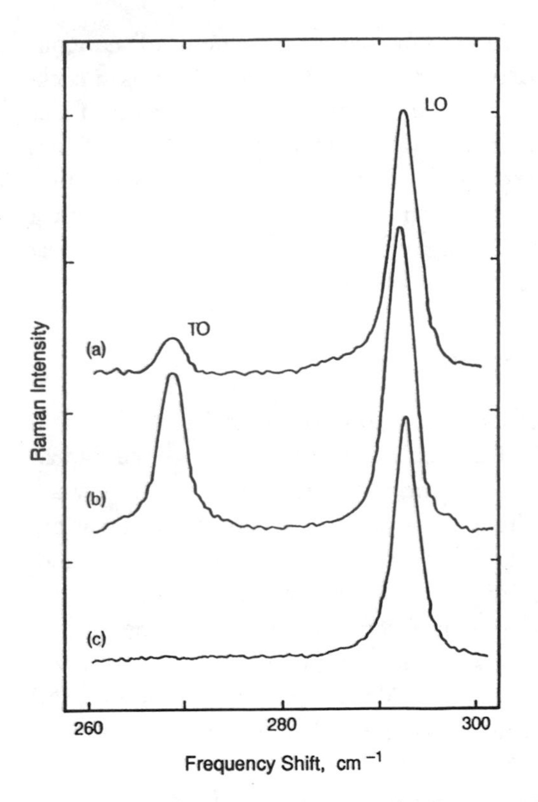

Figure 3.1 **First-order Raman scattering spectra comparing (*a*) GaAs/DSPE-SOS, (*b*) GaAs/CVD-SOS, and (*c*) homoepitaxial GaAs (100). The LO phonon modes are shifted to lower frequences with respect to the frequency of homoepitaxial GaAs (100). The shift is indicative of tensile strain present in the GaAs layers. (From Reference 18.)**

on silicon and silicon-on-sapphire substrates. Shown in Figure 3.1 are Raman spectra obtained from heteroepitaxial GaAs (100) layers deposited on commerically available chemical-vapor-deposited (CVD) silicon-on-sapphire (SOS) and SOS substrates that have been upgraded by the double solid-phase epitaxial (DSPE) process. It is apparent that both the LO and TO phonon modes are observed in the grown layers. Furthermore, an inspection of the grown films by TEM has revealed the presence of a high density of threading dislocations and microtwins. However, similar dislocation densities have been observed in both GaAs/SOS samples. As verified by TEM, the microtwin density at the top of the films is lowest in those GaAs epitaxial films grown on DSPE-SOS substrates. As a consequence, the intensity of the corresponding TO mode is significantly reduced, as shown by Figure 3.1*a*.

Raman scattering studies pertaining to the lattice dynamics of III–V ternary alloys have been extensively reported.[13, 14, 19–28] For example, the III–V semiconductor alloys, namely, AlGaAs and InGaAs systems, have received the most attention because of their continuous miscibility and abilitity to form solid solutions

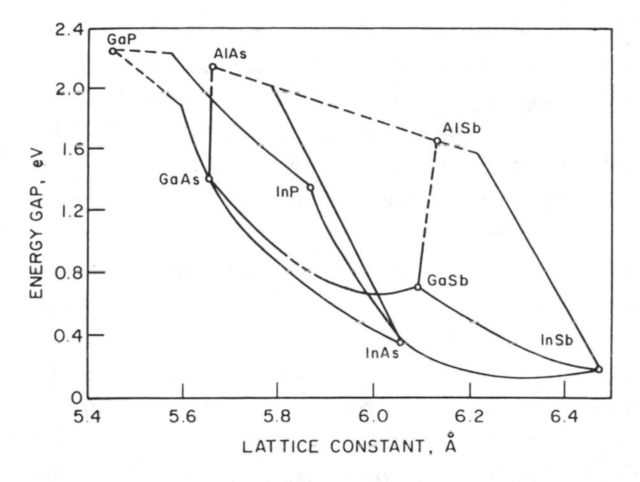

Figure 3.2 **Energy bandgap and lattice constant for various III–V semiconducting compounds. (From Reference 30.)**

with a desired alloy composition.[29] Consequently, the lattice constant of these III–V alloys can be easily varied and tailored for specific electronic device and optoelectronic applications. Figure 3.2[30] illustrates the variation of bandgap with lattice-constant for the common III–V compounds and their ternary derivatives.

With the exception of GaInP, which exhibits an apparent "one-mode" behavior, most III–V alloy semiconductors display what is called a "two-mode" frequency dependence throughout the entire alloy composition range. The underlying physical aspects of predicting the corresponding mode behavior in ternary alloys have been reviewed extensively by Landa et al.[31] In the case of $Al_xGa_{1-x}As$ ($0 < x < 1$), a "two-mode" behavior is displayed with the presence of two pairs of LO and TO modes, both GaAs-like and the AlAs-like. The corresponding Raman frequency dependencies for the (100) face of $Al_xGa_{1-x}As$ are illustrated in Figure 3.3, where the frequencies of the AlAs-like modes, designated LO_1 and TO_1, and the GaAs-like

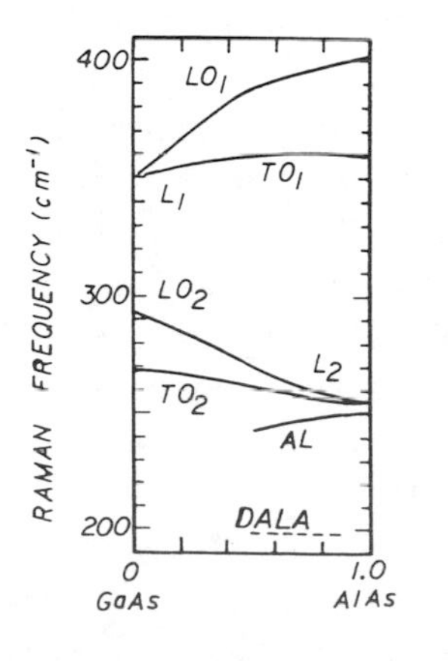

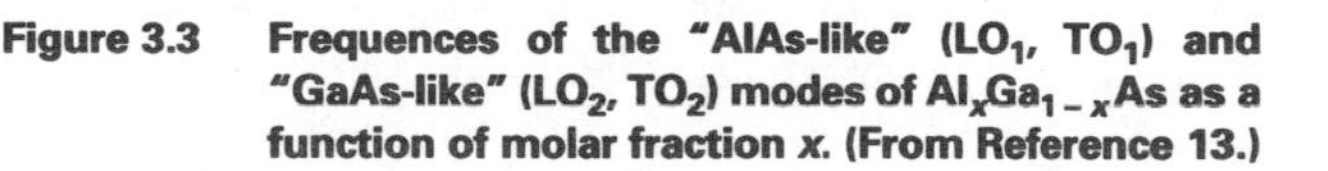
Figure 3.3 **Frequences of the "AlAs-like" (LO_1, TO_1) and "GaAs-like" (LO_2, TO_2) modes of $Al_xGa_{1-x}As$ as a function of molar fraction x. (From Reference 13.)**

modes, designated LO_2 and TO_2, are plotted as a function of composition, x. It is apparent that the frequency dependence of these modes provides a convenient and nondestructive method for establishing the alloy composition, x. Moreover, since Raman scattering can provide information on a local scale (order of several lattice constants), it is possible to probe the structural or topological disorder in the as-grown epitaxial alloys. In a III–V ternary semiconductor, there exist alloy potential fluctuations due to the microscopic nature of the compositional disorder. In consequence, translational invariance is broken, resulting in broadening and asymmetry of the LO Raman line shapes. This effect will manifest itself as a breakdown of the q-vector selection rule. For example, in $Al_xGa_{1-x}As$ films both the LO_1 and LO_2 phonon modes are asymmetric, with a total line width greater than that of the binary endpoint materials. Indeed, a theoretical interpretation of the corresponding Raman line shapes for $Al_xGa_{1-x}As$ has been presented by Parayanthal and Pollak[32] using an alloy potential fluctuation model based on q-vector relaxation. Further evidence pertaining to local composition disorder has also been observed in the epitaxial growth of $GaP_{1-x}Sb_x$ and $GaAs_{1-x}Sb_x$ ($0 < x < 1$) ternary alloys.[23, 28] In particular, for $GaP_{1-x}Sb_x$ a "two-mode" behavior is reported over the entire composition range, as illustrated in Figure 3.4. Moreover, it has also been shown that the intensities of the GaP (LO1 = 403 cm^{-1}) and GaSb (LO2 = 237 cm^{-1}) phonon modes are approximately proportional to the molar fractions x and $1 - x$, respectively. The increase in spectral broadening and asymmetry of the GaP (LO_1) phonon mode with increasing antimony content is noteworthy. In particular, this observation has been interpreted in terms of compositional fluctuations in the alloy occurring from spinodal decomposition. In consequence, these compositional fluctuations lead to an increase in phonon scattering and relaxation in the q-vector selection rule, resulting in a broadening in the LO_1 and LO_2 Raman peaks on their low-energy side. Indeed, confirmation of spinodal decomposition has been obtained by TEM examination of the growth layers. Furthermore, it is noteworthy that similar results have also been reported for the $GaAs_{1-x}Sb_x$ material system.[28]

Recently, the $InAs_{1-x}Sb_x$ alloy semiconductor system has been investigated as the smallest energy bandgap ternary alloy for the potential fabrication of infrared sources and detectors operating in the 8–12-μm spectral range. The first Raman study of $InAs_{1-x}Sb_x$ alloys grown by metal organic chemical vapor deposition (MOCVD) was performed by Cherng et al.,[25] who reported "mixed-modes." In particular, they observed a switch from one-mode behavior for compositions of $x < 0.6$ to a two-mode behavior for larger values of x. In contrast, Li et al.[26] have reported a two-mode behavior which is clearly visible over the entire alloy range in samples grown above 400 °C by molecular beam epitaxy (MBE). Furthermore, it has also been shown that the frequency-composition relation determined from the dominant InAs-like LO phonon in the homogeneous films is linear. However, for those epilayers grown below 400 °C there is a significant shift in the InAs-like LO phonon frequency which is believed to be indicative of phase separation. Indeed, a

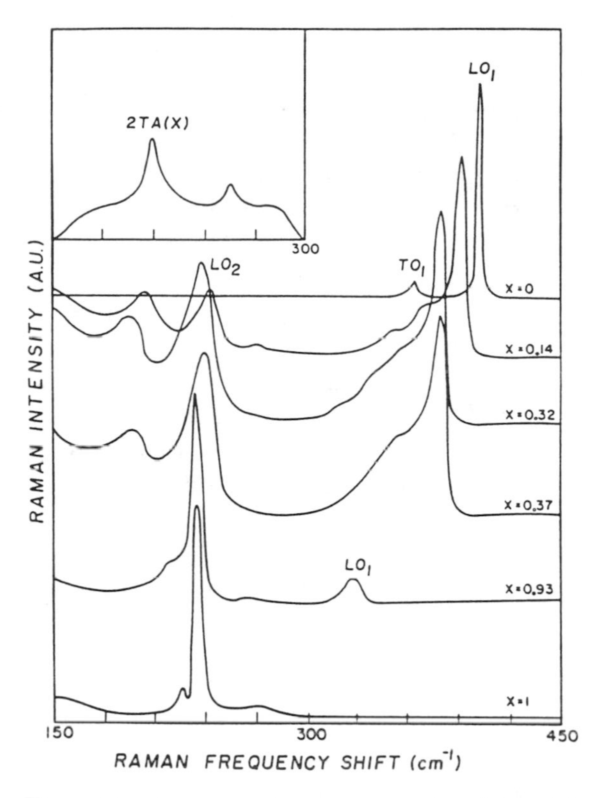

Figure 3.4 Raman scattering spectra for $GaP_{1-x}Sb_x$. The inset shows a broad band between 200 and 220 cm^{-1}, which may be due to a GaP-like 2TA(X) mode. (From Reference 23.)

direct comparsion with TEM results shows the formation of a superlattice structure composed of $InAs_{0.33}Sb_{0.67}$–$InAs_{0.69}Sb_{0.31}$.

Of current technological and scientific interest is the growth and characterization of the direct wide bandgap III–V nitrides, namely, GaN (E_g = 3.4 eV) and AlN (E_g = 6.2eV).[33] In particular these III–V binary systems are considered potential material candidates for the fabrication of optical devices which are active at wavelengths from the blue to well into the ultraviolet. At present there is a major effort to grow stoichiometric, high-quality heteroepitaxial films of these materials. In order to identify the crystallographic phases of the deposited layers, researchers employ the techniques of X-ray diffraction (XRD), TEM, and Raman spectroscopy widely. However, since Raman scattering is nondestructive and sensitive to a few hundred Ångstroms of material, the technique is extemely well-suited to the crystallographic phase analysis of thin single-crystal films. Recently, Raman scattering has been employed to identify both the wurtzitic (hexagonal) and zincblende (cubic) single-crystal phases of epitaxial GaN films grown on (0001) sapphire substrates.[6]

3.4 Group IV Materials

Recent advances in epitaxial thin-film growth processes have demonstrated high-quality films of Si_xGe_{1-x} for potential opto-electronic applications.[34] The alloys are often considered both for opto-electronic applications or for substrate engineering based on variations of the lattice constant. The binary phase diagram shows a continuous solid solubility of the alloy for all compositions. Issues of stoichiometry will relate to doping and vacancy formation.

An important aspect of the alloy system is that the bandgap can be varied with alloy composition. The results are summarized in Figure 3.5.[35, 36] The variation of the band gap (at 90 K) of the bulk alloys decreases gradually from ~1.15 to ~0.95eV as the alloy composition is varied from pure Si to $Si_{0.15}Ge_{0.85}$. For alloy concentrations in the range $0.85 < (1 - x) < 1$, a more rapid decrease in the band gap is observed. This difference in the dependence of the band gap trends is observed because the minimum of the conduction band in the alloys for $(1 - x)$ is along the Δ direction (i.e., along the [100] direction) in the Brillouin zone, whereas for the Ge-rich alloys, the minimum in the conduction band is at the L point (along

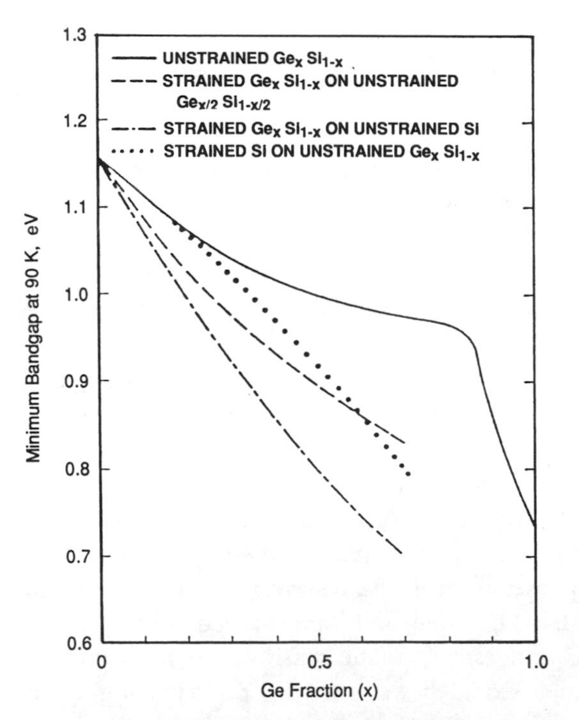

Figure 3.5 **The variation of the bandgap as a function of Ge concentration in SiGe alloys. The upper curve is the bandgap of SiGe alloys. The other curves represent experimental and theoretical calculations for different strained configurations as indicated. (The composite figure is from Reference 36.)**

the [111] direction). For all the alloys, the maximum of the valence band is at the Γ point.

Although the variation of the bulk bandgap itself offers significant opportunity for optical characterization, an additional factor that must be accounted for in epitaxial thin films is strain. The strain results from the lattice mismatch of the film and substrate. The lattice constant of the SiGe alloys ranges from 5.43 Å for pure Si to 5.66 Å for pure Ge. Under some growth conditions, the strain can be at least partially relaxed by the formation of misfit dislocations at the interface. For this particular alloy system, the strain causes substantial changes in the bandgap of the alloys.[35–37] Examples of the results are also summarized in Figure 3.5. These results have been both deduced theoretically and measured experimentally.

Thus, to characterize the optical properties of thin films of these alloys it is necessary to determine both the alloy composition and the intrinsic strain present in the film. Raman spectroscopy has proved useful in this case. An example of Raman spectra of strained and strain relaxed films is shown in Figure 3.6.[38] The Si_xGe_{1-x} alloys yield three relatively sharp features at ~500, ~400, and ~300 cm^{-1}

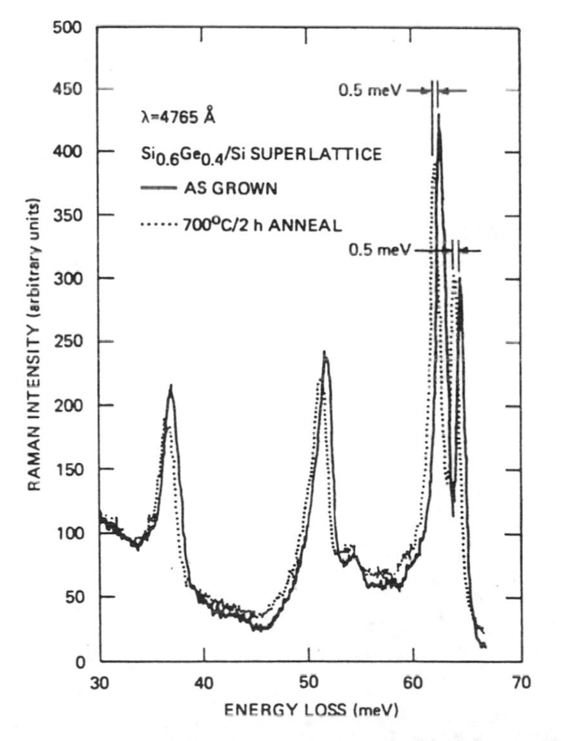

Figure 3.6 **Raman spectra of a pseudomorphic $Si_{0.6}Ge_{0.4}$/Si supperlattice as grown and after annealing for strain relaxation. The peaks near 62, 52, and 37 meV are attributed to the Si–Si, Si–Ge, and Ge–Ge vibrations in the SiGe layers, and the feature at 64.5 meV is due to the Si layers (1 meV = 8.066 cm^{-1}). (From Reference 38.)**

which have been assigned to Si–Si, Si–Ge, and Ge–Ge vibrations. There have been several studies of the peak frequency and linewidth variations for the alloys, and the results of the peak frequencies are summarized in Figure 3.7.[39–41] Because the ~500 cm^{-1} feature is strongest for Si-rich alloys and the ~300 cm^{-1} feature is strongest for Ge-rich alloys, these two features are most often used to determine the alloy concentration. It has also been shown that the relative intensities of the peaks can be used to determine the alloy concentration.[39] The predicted ratios of Ge–Ge:Ge–Si:S–Si should follow $(1 - x)^2:2x(1 - x):x^2$. A recent analysis has explored in some detail the variations in previous observations and explored methods to quantify the analysis.[40]

The strain in the films can be controlled several ways. For low-temperature growth with a film thickness below the critical thickness, the layer will be pseudomorphically strained to the substrate. The strain can be relaxed either by growth at higher temperatures or postdeposition annealing or by growth above the critical thickness. The strain is relaxed by the production of misfit dislocations at the interface. The optical properties of strained SiGe alloys on Si are quite different from those of the unstrained alloys. The results of the calculation by People and Bean[35, 36] is shown in Figure 3.5. Similar results have been obtained by Van de Walle and Martin, and experimental results are in good agreement with the calculations.[37] An example of how the strain can be detected in the Raman measurements is shown in Figure 3.6.

We note that many different strained systems are possible. Another possibility would be to grow Si on a SiGe alloy substrate. The Si would then be strained to exhibit a pseudomorphic interface with the alloy. The predicted optical responses of these structures are also summarized in Figure 3.5.

It is worthwhile mentioning a natural extension of this analysis into superlattices of Si and Ge.[42] These structures would be strained to match the substrate. A novel approach to distributing the strain in the superlattice is to grow the superlattice on

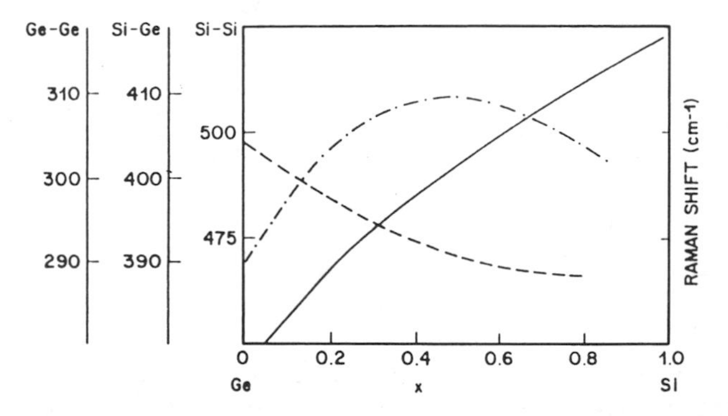

Figure 3.7 **The alloy dependence of the peak frequencies of the Si–Si, Si–Ge, and Ge–Ge Raman features for SiGe alloys. (Data is from Reference 39, and the composite figure is from Reference 41.)**

a $Si_{0.5}Ge_{0.5}$ substrate.[42] In this way, both the Si and Ge layers would be strained (with Si tensile and Ge compressive), the critical thickness of the superlattice layers would be substantially increased, and the whole superlattice would not have a critical thickness. This structure has been termed a strain symmetrized superlattice. An example involving the Raman analysis of a strain symmetrized superlattice is shown in Figure 3.8.[43] The results indicate very weak scattering from SiGe modes, which in turn indicates sharp interfaces. In addition, the shifts of the peaks verify that the Si and Ge layers are strained in the directions expected. Other effects due to confinement of the phonons must also be considered to describe the peak positions accurately.[43]

3.5 Amorphous and Microcrystalline Semiconductors

In general, Raman scattering conserves momentum selection rules. Since a wavevector cannot be related to the vibrational modes of amorphous materials, all the vibrational modes will have a component of the wavevector which satisfies the momentum conservation conditions.[44] These conditions are of course determined by the scattering geometry and the wavelength of the incident and scattered light. Thus, the Raman scattering spectrum often resembles the density of vibrational states. The Stokes shifted intensity, $I_s(\omega)$, is often written as

$$I_s(\omega) = \Sigma_b C_b(\omega) G_b(\omega) \frac{n(\omega)+1}{\omega} \tag{3.1}$$

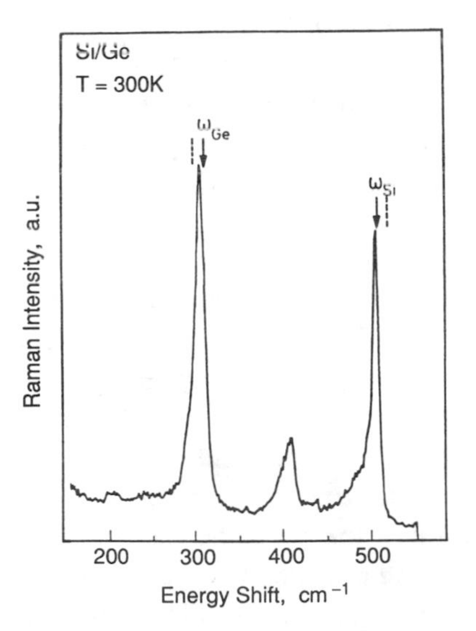

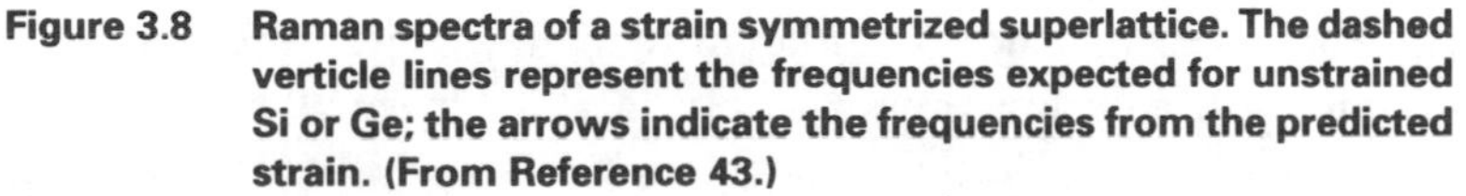

Figure 3.8 **Raman spectra of a strain symmetrized superlattice. The dashed verticle lines represent the frequencies expected for unstrained Si or Ge; the arrows indicate the frequencies from the predicted strain. (From Reference 43.)**

where $C_b(\omega)$ is the Raman matrix element, $G_b(\omega)$ is the density of states, and $n(\omega)$ is the boson occupation factor.[44] The subscript b refers to a vibrational band, and the sum is over all of the bands. Here it is assumed that the Raman matrix element can vary significantly for each band (in both intensity and polarization selection rules), but that the matrix element is relatively constant within a band.

The Raman scattering from microcrystalline semiconductors exhibits partial relaxation of the wavevector (or momentum) conservation selection rules. By considerations from the Heisenberg uncertainty principal, the wavevector uncertainty, Δk, would be

$$\Delta k = 2\pi / L \tag{3.2}$$

where L is the crystalline domain size. The theory has been worked out in detail for general materials[45, 46] and for Si in particular.[47] The results for Si have shown that the TO peak frequency and lineshape (or linewidth) can be related to the crystalline domain size. It has also recently been shown that the same relationship may be used for diamond because of similarities in the phonon dispersion of the corresponding TO modes of the two materials.[48]

Chalcogenide Glasses

A group of glasses based on S and Se are often classified as the semiconducting chalcogenide glasses. As, Ge, and Si chalcogenide glasses can be formed from melts over relatively large alloy ranges, and alloys of these compounds are also obtainable. The materials can also be produced as thin films using various deposition processes. The materials have proven useful for xerographic applications.

An example of the dependence of the optical response versus alloy concentration for several binary alloys is shown in Figure 3.9.[49] Note that a minimum or maximum in the optical bandgap is observed at the chemically ordered composition (i.e., $GeSe_2$, As_2S_3, or As_2Se_3).

One of the most important methods of characterizing these amorphous semiconductors has been Raman scattering. Unlike the Raman scattering from group IV amorphous semiconductors, these materials often show relatively sharp bands with strong polarization characteristics. These observations have led investigators to propose a model based on molecular structures to explain the sharpness, polarization dependence, and IR-Raman activity of various modes.[50] Although it was never envisaged that the glasses were actually composed of molecules, the model has been used to provide a simple method for characterizing the vibrational properties. The model has proven particularly successful in the S and Se chalcogenide glasses, and this has been explained by noting that the near 90° bond angle at the chalcogenide site tends to decouple the stretching vibrations of the different bonds.

In amorphous materials it may not be apparent that alloy concentration plays a key role in determining the optical properties. In fact, if binary alloys were absolutely random, a wide range of different local atomic arrangements would be

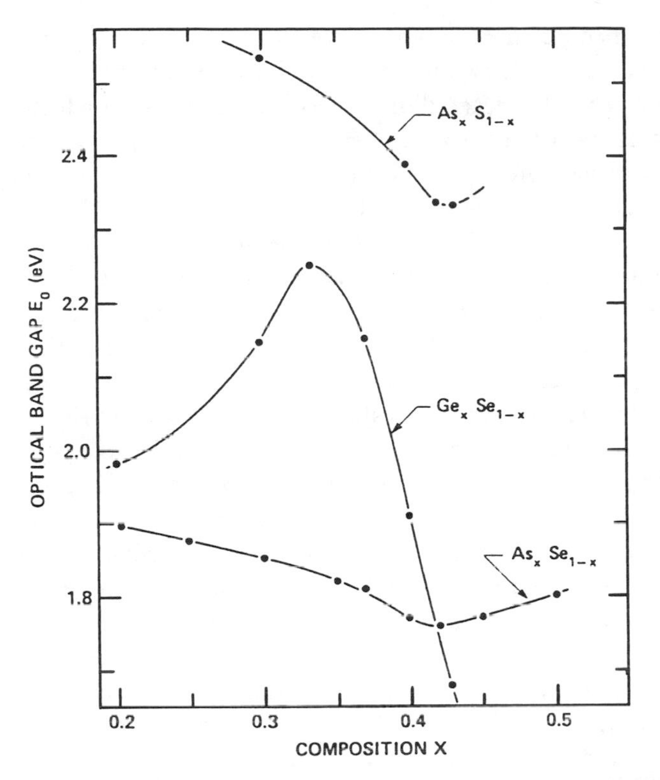

Figure 3.9 Room-temperature optical bandgap for bulk glasses in the Ge_xSe_{1-x}, As_xS_{1-x}, and As_xSe_{1-x} alloy systems. (From Reference 49.)

possible. It has been shown, however, that covalent bonding exists, and a strong trend towards chemical ordering often dominates the possible configurations.[50, 51] The model has been termed the chemically ordered covalent random network (COCRN). The basic aspects are that the bonding follows the 8-*n* rule with group IV, V, and VI atoms coordinated fourfold, threefold, and twofold, respectively. In addition, heteropolar bonding (i.e., chemical ordering) occurs when possible.

The essentials of this model were verified for bulk glasses in several alloy systems, including Ge_xS_{1-x},[51] Ge_xSe_{1-x},[52, 53] As_xS_{1-x},[54] and As_xSe_{1-x}.[54] In these studies, the Raman spectra of the glasses were obtained and analyzed. Following the model described above, chemical ordering should occur at compositions $Ge_{0.33}S_{0.67}$ (or GeS_2) and $As_{0.40}S_{0.60}$ (or As_2S_3). At these compositions, features associated with only Ge–S or As–S bonding were observed. For compositions to the S-rich side of this composition, additional features associated with S–S bonding were observed. Similarly, for compositions to the Ge- or As-rich side, features associated with Ge–Ge- or As–As-containing structures were observed in addition to the heteropolar bonding. The combination of the optical and structural studies showed that ordered glasses could be produced with variable optical bandgaps.

In essence, the previously mentioned studies could be viewed as measuring the alloy composition of the continuous binary alloy. Stoichiometric deviations in this sense apparently do not make sense. However, if the network exhibits defects which break the "rules" of the chemically ordered covalent network, then these cases could be viewed as deviations from stoichiometry. Such is the case for evaporated films of these glasses. The Raman spectra for films evaporated from As_2S_3 glasses are shown in Figure 3.10.[55] The as-deposited film shows many sharp spectral lines which were not evident in the spectrum of the bulk As_2S_3 glass. In fact, lines are observed which were attributed to structures containing both As–As and S–S lines. The peak structure at ~500 cm^{-1} was attributed to S–S bonds, whereas the series of sharp bands extending from 100 to 370 cm^{-1} correspond to As_4S_4 molecule vibrations, and these structures contain As–As bonding. After annealing at near the glass transition temperature, the S–S feature disappears, whereas the features of the As_4S_4 molecules are substantially reduced. Apparently, the annealing resulted in "polymerization" of the S–S and As–As bonds. The "wrong bonds" in the as-deposited films indicate that the COCRN model does not apply directly. The results were analyzed in terms of a completely random covalent network model and a model where ordered structures were preserved. Here it was found that the as-deposited As chalcogenides tended towards the ordered structures, and this was accounted for because the deposition vapor contained molecules with these structures (i.e., As_4S_4 which contains S_2–As–As–S_2). In contrast, the "wrong bonds" or stoichiometry deviations in Ge–Se films indicated that a random configuration applied.[55]

The Raman spectra were also used to characterize the films after annealing.[55] For the films formed from thermal evaporation of As_2S_3, the resulting spectra after annealing were similar to those of the bulk glass with a concentration of $As_{0.42}S_{0.58}$. Thus, the technique has identified the alloy concentration, and the degree of "wrong bonds" in the as-deposited films can be considered deviations from stoichiometry.

An unusual aspect of the microstructure of the chemically ordered films has been the subject of many studies.[52, 53, 56, 57] The Raman scattering results for Ge_xSe_{1-x} are shown in Figure 3.11.[57] Stretching vibrations associated with Ge–Se, Se–Se, and Ge–Ge bonding structures were identified at ~200 cm^{-1}, 250 cm^{-1}, and 185 cm^{-1}, respectively.[52, 53, 56, 57] All three features exhibited strong polarization characteristics with a HH-to-VH ratio of as large as 20.[53] It was found that the feature associated with Se–Se bonding was only observed for $x < 0.33$. Similarly, the Ge–Ge feature was observed for $x > 0.33$, again consistent with the COCRN model. One of the strongest features in the spectrum was not accounted for by this model. The feature at ~220 cm^{-1} was shown to have a maximum intensity at the chemically ordered composition ($GeSe_2$) which then decreased rapidly for deviations from this composition. Several of the atomic models which were initially suggested to explain this feature include (1) coupling between molecular units,[52] (2) large ring structures,[53] (3) large planar structures terminated with Se–Se bonds,[56] and more recently, (4) small rings or corner-connected tetrahedra.[58, 59]

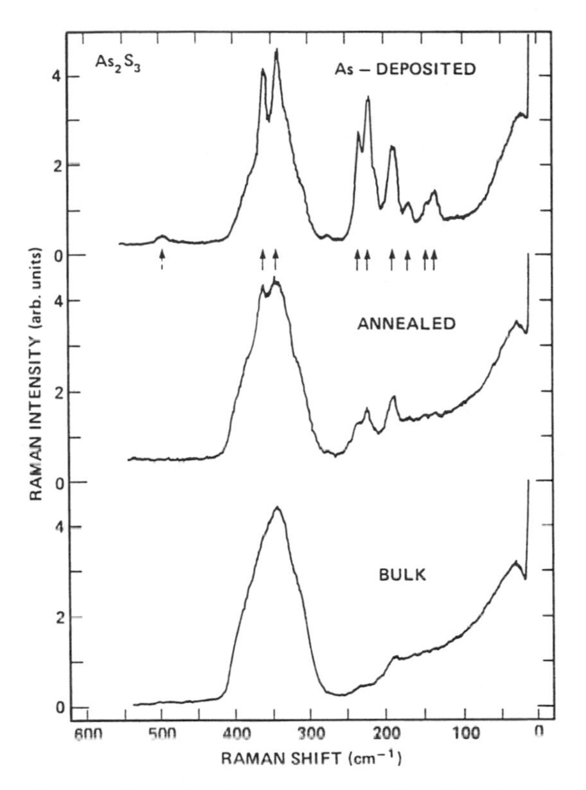

Figure 3.10 **The Raman spectra of a film deposited from evaporation of As_2S_3 as-deposited and after annealing at near the glass transition temperature, and the bulk As_2S_3 glass. The arrows at Raman shifts less than 400 cm^{-1} are attributed to structures containing As–As bonds, whereas the feature at ~500 cm^{-1} is due to S–S bonds. (From Reference 55.)**

The recent work by Sugai[57] has strongly supported the presence of small rings characterized by edge-sharing tetrahedra. The work has included an extensive analysis of Ge and Si chalcogenide glasses, and a stochastic random covalent network has been proposed to describe the bonding properties leading to the ring structures. Although the current understanding of the structures seems consistent, there has not been a direct correlation with the optical properties of the materials.

Group IV Microcrystalline Semiconductors

A potential for large-area thin film opto-electronics includes amorphous and microcrystalline semiconductors.[60] There has been a substantial amount of research into the properties of films of Si which exhibit crystalline regions. Although these films which were microscopically crystalline were termed microcrystalline, this is probably a misnomer, since the regions of the film which exhibit crystalline order have domain sizes as small as a few nanometers. Thus it may be more appropriate to term

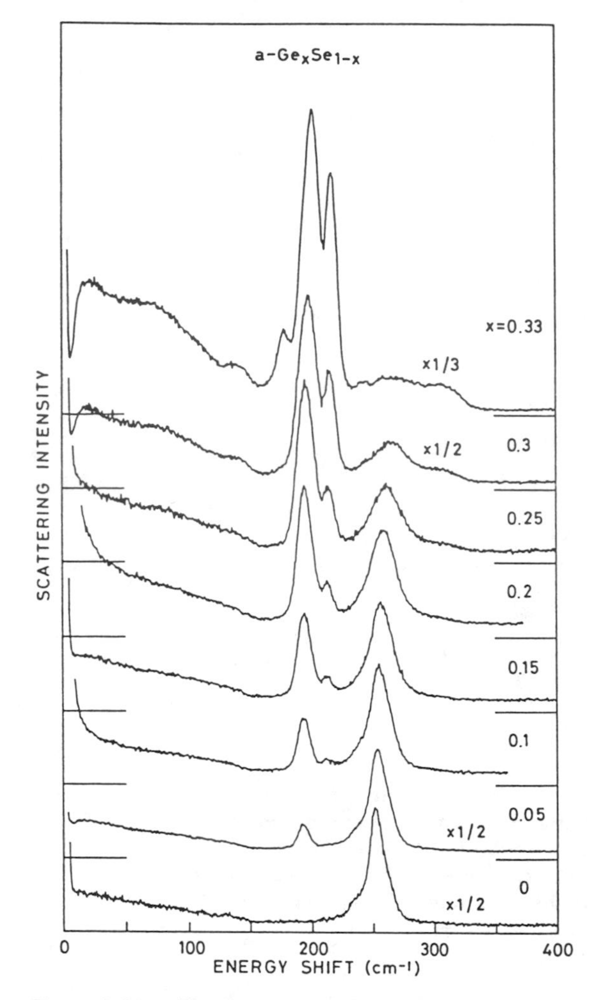

Figure 3.11 **The Raman spectra of GeSe glasses. The features at 200 and 250 cm^{-1} are attributed to $GeSe_4$ and Se–Se structures, respectively, but the feature at ~220 cm^{-1} has been associated with other network structures. (From Reference 57.)**

these films as nanocrystalline. Since the films are prepared by the same plasma CVD methods used to obtain amorphous films, they are often compared with the amorphous counterparts, which for Si would be hydrogenated amorphous Si (a-Si:H).[61] With the advancement of high-temperature plasma CVD systems, diamond films have also been produced. Several similarities should be considered in the characterization of films of these two materials. These are related to the facts that the materials are actually composites and that the optical properties of the different materials vary substantially. In the case of Si, the composite is crystalline Si and a-Si:H, whereas for the diamond films, the components are diamond and disordered sp^2 graphitic regions.[62] In both cases, the disordered regions exhibit substantially increased optical absorption over the ordered regions. Within the context of this review on composition and stoichiometry analysis, the two different components can be considered the components of an alloy. Clearly, the optical properties are strongly dependent on the relative amounts of the components.

The optical absorption differences of the domains will affect the observed Raman results.[63, 63] The ratio of the Raman intensities of the transparent and absorbing regions can be given by

$$\frac{I_a}{I_t} = \frac{A_a N_a V_a}{A_t N_t V_t} \tag{3.3}$$

where *I*, *A*, *N*, and *V* are the Raman intensity, Raman cross section, atomic density, and illuminated volume, respectively, and the subscripts *a* and *t* refer to the absorbing and transparent components. For the transparent regions, the illuminated volume will depend on the relative fraction of the transparent regions, $(1 - P_a)$. In contrast, the illuminated volume of the absorbing regions will depend on the relative fraction of the absorbing material, P_a, and on the absorption constant, α_a, and size of the domains of the absorbing regions, l. The key aspect of the analysis was to account for the fact that the absorbing regions would not be sampled completely. Equation 3.3 then becomes

$$\frac{I_a}{I_t} = \frac{A_a N_a P_a}{A_t N_t (1 - P_a)} \cdot \frac{1}{\alpha_a l} \tag{3.4}$$

This equation is valid for $\alpha_a^{-1} < l$. This model, which neglects effects due to field enhancement and different shapes of the absorbing regions, model was calculated using optical constants for diamond/graphite and crystalline Si/a-Si:H.[63] The results for Si are shown in Figure 3.12. For the calculation, the ratio of the Raman

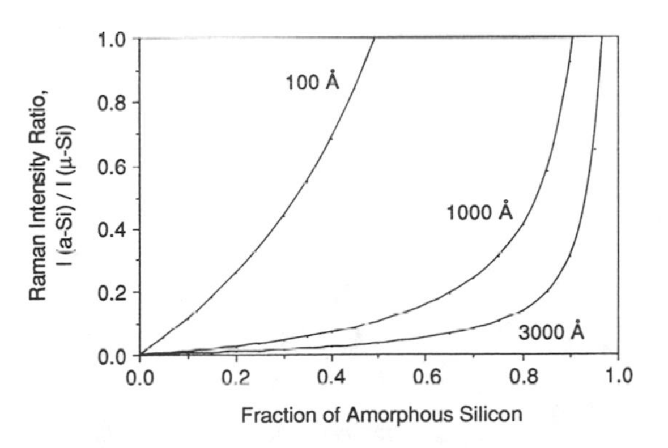

Figure 3.12 **The theoretical ratio of the Raman signal from the amorphous and crystalline regions of a microcrystalline Si film versus the fraction of amorphous region. Note that the calculation would indicate different observations if the domains of the absorbing regions were different. (From Reference 63.)**

cross sections was determined from the peak intensities of the materials, α for a-Si was taken to be 2×10^5 cm^{-1}, and the densities were assumed to be the same. From Figure 3.12 it is clear that the observed Raman ratio cannot be directly related to the amorphous/microcrystalline ratio unless the domain size is known. However, below a domain size of about 500 Å, the model will break down, and in the limit of very small domains, the term $1/\alpha_a l$ will go to 1. In that limit, the spectra can be decomposed directly into amorphous and crystalline components without accounting for the variations in the absorption of the different regions.

An example of Raman scattering from microcrystalline Si films produced by plasma CVD techniques is shown in Figure 3.13.[63] From comparison with previous results, it appears a reasonable approximation that both the absorbing and transparent regions exhibit domains less than 500 Å. Thus, the spectrum was analyzed by decomposing into amorphous and microcrystalline components. In the case shown, the analysis indicated that the film comprised 78% amorphous and 22% microcrystalline components.

3.6 Summary

The discussion in this review has emphasized alloy composition and stoichiometry measurements for thin films with potential optical or opto-electronic applications. The materials described exhibit significant variations in optical response for the alloy composition or stoichiometry in the case of amorphous materials.

The discussion has focused on the analysis of the vibrational modes detected by Raman spectroscopy. In most of the examples described in this review, the observations have involved the strongest features in the spectra, rather than relying on weakly observable features. There have been numerous studies of weaker features

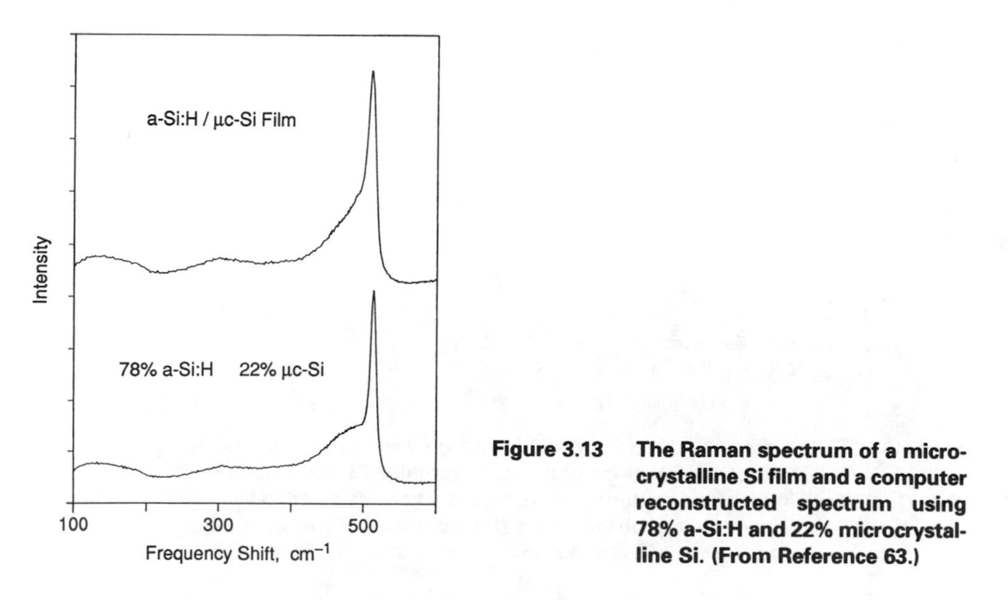

Figure 3.13 **The Raman spectrum of a microcrystalline Si film and a computer reconstructed spectrum using 78% a-Si:H and 22% microcrystalline Si. (From Reference 63.)**

which are often used to relate specific defect or impurity structures. The analysis of the strong features has involved using the central peak frequency, the linewidth, or the integrated intensity to deduce the alloy composition or stoichiometry. Recent advances in Raman scattering instrumentation, including liquid-nitrogen-cooled charge coupled device (CCD) detectors and appropriate spectrometers, allow for rapid measurement of these features. Software developments have also allowed rapid analysis of the peak frequency, linewidth, or integrated intensity. With these advances it is often possible to measure alloy concentration to ±2%. Relative concentrations between similar samples can be measured even more accurately.

The experiments described, in general, involved above bandgap laser excitation. The laser line wavelength could be chosen to match the absorption depth of the area to be studied. In some cases, an appropriate choice of laser line could lead to composition determination of buried layers. Most of the experiments described here were carried out with ion laser excitation (either Ar or Kr). Recent advances in laser technology will allow more flexibility in the choice of laser wavelength and, again, an increased potential in the ability to tune the measurement to examine a particular film in a multilayer structure. One complication (that can prove an advantage in some applications) is that interference properties in a multilayer structure can strongly affect the light-intensity distribution in the film.

Over the last two decades there have been substantial efforts at measuring and understanding the optical properties and the Raman scattering from optical materials. These measurements coupled with new advances in the measurement techniques now allow Raman scattering as a rapid nondestructive tool for determining thin film compositions, stoichiometry, and related microstructures. As new materials such as wide bandgap semiconductors are more thoroughly studied, we envision continued new applications of the techniques. One area that could prove particularly exciting is the potential for using these optical techniques in situ in a growth environment. Many experiments are already addressing this potential application for particular materials systems.

Acknowledgment

We acknowledge the research support from the following organizations: the National Science Foundation, the Office of Naval Research, Kobe Research and Development, and the Research Triangle Institute.

References

1 *Non-Stoichiometry in Semiconductors.* (K. J. Bachman, H.-L. Hwang, and C. Schwab, Eds.) North-Holland, Amsterdam, 1992.

2 W. Hayes and R. Loudon. *Scattering of Light by Crystals.* John Wiley and Sons, New York, 1978.

3 *Light Scattering in Solids,* Vol. 1. (M. Cardona, Ed.) Springer-Verlag, Berlin, 1983.

4 F. H. Pollak. In *Analytical Raman Spectroscopy.* (J. G. Grasselli and B. J. Bulkin, Eds.) John Wiley & Sons, New York, 1991, p. 137.

5 A. S. Barker, Jr., and A.J. Sievers. *Reviews of Modern Physics.* **47** (Suppl. No. 2), S1, 1975.

6 T. P. Humphreys, C. A. Sukow, R. J. Nemanich, J. B. Posthill, R. A. Rudder, and R. J. Markunas. *Mater. Res. Soc. Symp. Proc.* **162**, 531, 1990.

7 K. Kubota, Y. Kobayashi, and K. Fujimoto. *J. Appl. Phys.* **66**, 2984, 1989.

8 S. Strite, J. Ruan, Z. Li, N. Manning, A. Salvador, H. Chen, D. J. Smith, W. J. Choyke, and H. Morkoc. *J. Vac Sci. Technol.* **B9**, 1924, 1991.

9 G. Abstreiter, R. Trommer, M. Cardona, and A. Pinczuk. *Solid State Commun.* **30**, 703, 1979.

10 A. Mooradian and A. L. McWhorter. *Phys. Rev. Lett.* **19**, 849, 1967.

11 R. Trommer, H. Muller, M. Cardona, and P. Vogl. *Phys. Rev.* **B21**, 4869, 1980.

12 E. Bedel, G. Landa, R. Charles, J. P. Redoules, and J. B. Renucci. *J. Phys.* **C19**, 1471, 1986.

13 R. Tsu. *Proc. SPIE–Int. Soc. Opt. Eng.* **276**, 78, 1981.

14 G. Abstreiter, E. Bauser, A. Fisher, and K. Ploog. *Appl. Phys.* **16**, 345, 1978.

15 J. Biellmann, B. Prevot, and C. Schwab. *J. Phys.* **C16**, 1135, 1983.

16 C. Fontaine, H. Benarfa, E. Bedel, A. Munoz-Yague, G. Landa, and R. Carles. *J. Appl. Phys.* **60**, 208, 1986.

17 G. Landa, R. Carles, J. B. Renucci, C. Fontaine, E. Bedel, and A. Munoz-Yague. *J. Appl. Phys.* **60**, 1025, 1986.

18 T. P. Humphreys, J. B. Posthill, K. Das, C. A. Sukow, R. J. Nemanich, N. R. Parikh, and A. Majeed. *Jpn. J. Appl. Phys.* **28**, L1595, 1989.

19 R. J. Nemanich, D. K. Biegelsen, B. A. Street, B. Downs, B. S. Krasor, and R. D. Yingling. *Mat. Res. Soc. Symp.* Vol. 116, 1988, p. 245.

20 N. Saint-Cricq, R. Carles, J. B. Renucci, A. Zwick, and M. A. Renucci. *Solid State Commun.* **39**, 1137, 1981.

21 K. Kakimoto and T. Katoda. *Appl. Phys. Lett.* **40**, 826, 1982.

22 Y. A. Aleshchenko, U. Zhumakulov, G. Ol'gart, and B. V. Baranov. *Sov. Phys. Doklady.* **27**, 480, 1982.

23 Y. T. Cherng, M. J. Jou, H. R. Jen, and G. B. Stringfellow. *J. Appl. Phys.* **63**, 5444, 1988.

24 T. C. McGlinn, T. N. Krabach, and M. V. Klein. *Phys. Rev.* **B33**, 8396, 1986.

25 Y. T. Cherng, K. Y. Ma, and G. B. Stringfellow. *Appl. Phys. Lett.* **53**, 886, 1988.

26 Y. B. Li, S. S. Dosanjh, I. T. Ferguson, A. G. Norman, A. G. de Oliveira, R. A. Stradling, and R. Zallen. *Semicond. Science and Technol.* **7**, 567, 1992.

27 B. Jusserand and J. Sapriel. *Phys. Rev.* **B24**, 7194, 1981.

28 R. M. Cohen, M. J. Cherng, R. E. Benner, and G. B. Stringfellow. *Appl. Phys.* **57**, 4817, 1985.

29 G. B. Stringfellow. *J. Cryst. Growth.* **58**, 194, 1982.

30 P. K. Bhattacharya and S. Dhar. In *Semiconductors and Semimetals, Vol. 26.* (R. K. Willardson and A. C. Beer, Eds.) Academic Press, New York, 1988, p. 147.

31 G. Landa, R. Chales, and J. B. Renucci. *Proceedings.* 18th Int. Conf. Phys. Semicond., Stockholm, 1987, p. 1361.

32 P. Parayanthal and F. H. Pollak. *Phys. Rev. Lett.* **52**, 1822, 1984.

33 R. F. Davis, Z. Sitar, B. E. Williams, H. S. Kong, H. J. Kim, J. W. Palmour, J. A. Edmond, J. Ryu, J. T. Glass, and C. H. Carter, Jr. *Mater. Sci. Eng.* **B1**, 77, 1988.

34 *Silicon Molecular Beam Epitaxy,* Vol. 2. (E. Kasper and J. C. Bean, Eds.) CRC Press, Boca Raton, FL, 1988.

35 R. People and J. C. Bean. *Appl. Phys. Lett.* **48**, 539, 1986.

36 J. C. Bean. *Mat. Res. Soc. Symp. Proc.* **116**, 479, 1988.

37 C. G. Van de Walle and R. M. Martin. *Phys. Rev.* **B34**, 5621, 1986.

38 R. J. Hauenstein, B. M. Clemens, R. H. Miles, O. J. Marsh, E. T. Croke, and T. C. McGill. *J. Vac. Sci. Technol.* **B7**, 767, 1989.

39 M. A. Renucci, J. B. Renucci, and M. Cardona. In *Proc. of the 2nd Conference on Light Scattering in Solids.* (by M. Balkanski, Ed.) Flammarion, Paris, 1971, p. 326.

40 I. P. Herman and F. Magnotta. *J. Appl. Phys.* **61**, 5118, 1987.

41 J. Menendez, A. Pinczuk, J. Bevk, and J. P. Mannaerts. *J. Vac. Sci. Technol.* **B6**, 1306, 1988.

42 H. Presting, H. Kibbel, M. Jaros, R. M. Turton, U. Menczigar, G. Abstreiter, and H. G. Grimmeiss. *Semicond. Sci. and Tech.* **17**, 1127, 1992.

43 E. Kasper, H. Kibbel, H. Jorke, H. Brugger, E. Friess, and G. Abstrieter. *Phys. Rev.* **B38**, 3599, 1988.

44 R. Shuker and R. W. Gammon. *Phys. Rev. Lett.* **25**, 222, 1970.

45 R. J. Nemanich, S. A. Solin, and R. M. Martin. *Phys. Rev.* **B23**, 6348, 1981.

46 Z. Iqbal and S. Veprek. *J. Phys.* **C15**, 377, 1982.

47 P. M. Fauchet and I. H. Campbell. *CRC Crit. Rev. Solid State and Mat. Sci.* **14**, S79, 1988.

48 Y. M. LeGrice, R. J. Nemanich, J. T. Glass, Y. H. Lee, R. A. Rudder, and R. J. Markunas. *Mat. Res. Soc. Symp. Proc.* **162**, 219, 1990.

49 R. A. Street, R. J. Nemanich, and G. A. N. Connell. *Phys. Rev.* **B18**, 6915, 1978.

50 G. Lucovsky and R. M. Martin. *J. Non-Cryst. Solids.* **8–10**, 185, 1972.

51 G. Lucovsky, F. L. Galeener, R. C. Keezer, R. H. Geils, and H. A. Six. *Phys. Rev.* **B10**, 5134, 1974.

52 P. Tronc, M. Bensoussan, A. Brenac, and C. Sebenne. *Phys. Rev.* **B8**, 5947, 1973.

53 R. J. Nemanich, S. A. Solin, and G. Lucovsky. *Solid State Commun.* **21**, 273, 1977.

54 G. Lucovsky, R. J. Nemanich, S. A. Solin, and R. C. Keezer. *Solid State Commun.* **17**, 1567, 1975.

55 R. J. Nemanich, G. A. N. Connell, T. M. Hayes, and R. A. Street. *Phys. Rev.* **B18**, 6900, 1978.

56 P. M. Bridenbaugh, G. P. Espinosa, J. E. Griffiths, J. C. Phillips, and J. P. Remeika. *Phys. Rev.* **B20**, 4140, 1979.

57 S. Sugai. *Phys. Rev.* **B35**, 1345, 1987.

58 R. J. Nemanich, F. L. Galeener, J. C. Mikkelsen, Jr., G. A. N. Connell, G. Etherington, A. C. Wright, and R. N. Sinclair. *Physica.* **117&118B**, 959, 1983.

59 G. Lucovsky, C. K. Wong, and W. B. Pollard. *J. Non-Cryst. Solids.* **59&60**, 839, 1983.

60 *Materials Issues in Microcrystalline Semiconductors.* (P. M. Fauchet, K. Tanaka, and C. C. Tsai, Eds.) Mat. Res. Soc. Symp. Proc., Vol. 164.

61 S. Veprek. *Mat. Res. Soc. Symp. Proc.* **164**, 39, 1990.

62 R. E. Shroder, R. J. Nemanich, and J. T. Glass, *Phys. Rev.* B

63 R. J. Nemanich, E. C. Buehler, Y. M. LeGrice, R. E. Shroder, G. N. Parsons, C. Wang, G. Lucovsky, and J. B. Boyce. *Mat. Res. Soc. Symp. Proc.* **164**, 265, 1990.

4

Diamond As an Optical Material

ALBERT FELDMAN, L. H. ROBINS, E. N. FARABAUGH,
and D. SHECHTMAN

Contents

4.1 Introduction

The high transmissivity of perfect diamond over extensive regions of the electromagnetic spectrum, in combination with diamond's great hardness, large abrasion resistance, high thermal conductivity, and chemical inertness, makes diamond a highly desirable optical material. Until recently, the high cost and small dimensions of optical-quality diamond have limited the use of diamond to a few specialized optical applications; the best optical-quality diamond, type IIa diamond, still comes from natural sources. However, new chemical vapor deposition (CVD) processes have the potential to make available inexpensive bulk diamonds of large dimensions and thin film diamond to cover large areas, thus making possible the widespread use of diamond optics and optical coatings.[1, 2, 3] X-ray windows of CVD diamond are already commercially available.[4] Anticipated optical applications include infrared windows and domes, high-power laser windows, and membranes for X-ray lithography.[5] Other possible applications include electroluminescent devices,[6] lasers,[7] and optical switches.[8]

This chapter focuses principally on the status of CVD diamond as an optically transparent material. CVD diamond, which is mainly polycrystalline, exhibits several materials problems that limit its optical transmission, such as scattering due to large surface roughness and absorption due to defects, nondiamond carbon phases, and impurities. New polishing methods have the potential for producing smooth surfaces in reasonable polishing times; however, work on surface figure, to our knowledge, has not yet been addressed. While intrinsic multiphonon processes limit the transmissivity of all diamond between 2.5 and 6.5 μm, CVD diamond usually contains high densities of lattice defects, causing normally forbidden single-phonon absorption processes to occur between 7.5 and 12 μm. In addition, free carrier absorption in CVD diamond has also been reported. While diamond windows less than 10 μm thick can be made transmissive to visible and ultraviolet radiation, thicker components scatter excessively and show absorption due to defects. Raman spectroscopy is the principal method for observing the presence of nondiamond carbon phases in CVD diamond.

An important method for examining point defects in CVD diamond is cathodoluminescence spectroscopy. When this method is conducted in a scanning electron microscope (SEM), cathodoluminescence images can be generated that provide information about the spatial distribution of luminescent defects on a submicrometer scale. However, the relationship between the observed cathodoluminescence and defect-induced optical absorption has not yet been established. Luminescence has been the basis for diamond devices that emit optical radiation. Lasers and electroluminescent devices have been made, but the quantum efficiency at present is too small for these devices to be of practical use.

A powerful method for observing defects close to the atomic level is high-resolution transmission electron microscopy. Here again, the relationship between observed defects and optical absorption has not been determined.

Continuing research is improving the optical quality of CVD diamond. Recently, novel methods for producing single-crystal (non-CVD methods) or near-single-crystal diamond have been reported. When fully developed, these methods could make available large optics of single-crystal diamond.

High deposition temperature, large thermal expansion mismatch, and poor adhesion now limit the use of CVD diamond as a hard coating material. To overcome some of these difficulties, researchers have developed a method to place a diamond coating on ZnSe by bonding the coating to the substrate with a chalcogenide glass layer.[9]

4.2 Deposition Methods

Several deposition methods are being used to deposit diamond by CVD. High-quality polycrystalline diamond has been made by hot filament CVD,[3] microwave plasma CVD,[10] dc plasma torch,[11] radio frequency plasma torch,[12, 13] microwave plasma torch,[14] and oxyacetylene torch.[15, 16] Figure 4.1 shows SEM images of

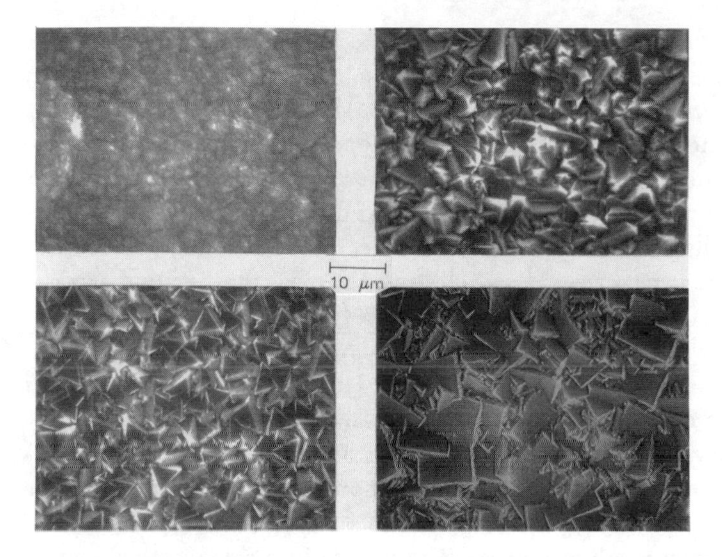

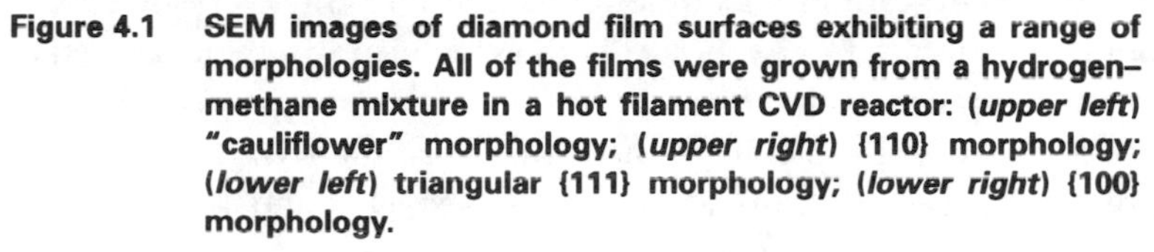

Figure 4.1 SEM images of diamond film surfaces exhibiting a range of morphologies. All of the films were grown from a hydrogen–methane mixture in a hot filament CVD reactor: (*upper left*) "cauliflower" morphology; (*upper right*) {110} morphology; (*lower left*) triangular {111} morphology; (*lower right*) {100} morphology.

diamond films showing typical surface morphologies.[17] The morphologies observed are common to all the deposition methods. The films were grown on silicon wafer substrates in a hot filament reactor under different deposition conditions.

Except for the oxyacetylene torch method, the charge gas for producing diamond is hydrogen mixed with a hydrocarbon such as methane, acetylene, methyl alcohol, ethyl alcohol, or acetone.[18] The hydrogen fraction in the feed gas is almost always greater than 90%. The quality of the diamond usually improves with increasing hydrogen fraction in the feed gas. It is also found that oxygen added to the charge gas improves the diamond quality.[19] Carbon monoxide mixed with hydrogen has also been used.[20] In the plasma torch methods, argon is sometimes used as a sheath gas. In the case of the oxyacetylene torch, only acetylene and oxygen comprise the charge gas. In this case, deposition is carried out in the reducing part of an oxygen-poor torch flame.

Substrate temperatures during deposition are usually maintained at a constant temperature in the range 600–1000 °C. Depositions have been done[21] at temperatures below 400 °C; however, the growth rates are low and the quality of the diamond, as determined by Raman spectroscopy, is poor. Recently, depositions by the oxyacetylene torch method at substrate temperatures up to 1400 °C have resulted in the growth of single-crystal diamonds at high growth rates.[22, 23] The method appears very promising for producing boule-sized single-crystal diamonds of high optical quality.

A creative method for producing nearly single-crystal diamond layers over large areas has been developed recently.[24] Selective area etching was used to produce an array of etch pits of a well-defined pyramidal shape in a single-crystal silicon substrate. A slurry containing single-crystal diamond particles of a matching pyramidal shape was applied to the surface of the substrate and then removed. Diamond particles that had been left behind were embedded in each of the etch pits. These particles were crystallographically oriented to within several degrees. When the substrate was placed in a CVD reactor, the diamond particles acted as seeds for diamond growth, resulting in a continuous diamond film that was nearly a single crystal.

There have been two reports of single-crystal diamond nucleation on single-crystal copper substrates. Both methods rely on the lack of solubility of carbon in copper. In one experiment,[25] carbon ions were implanted into a single-crystal copper substrate at an elevated temperature. A transmission electron microscope diffraction pattern indicated that a single-crystal diamond layer had formed on the surface of the copper.

In the other experiment, carbon ions were implanted into single-crystal copper substrates at room temperature.[26] The surface of the specimen was then exposed to high-power radiation from an excimer laser. The laser pulse energy was chosen to cause the surface of the copper substrate to melt and then to refreeze rapidly. During refreezing of the copper, the carbon atoms were expelled and forced to the surface, precipitating on the surface as a diamond layer.

A halogen-assisted CVD method has recently been discovered for producing diamond[27]; however, this research is in an early stage and claims for good-quality diamond films made by this method have not yet been reported.

4.3 Optical Properties of CVD Diamond

Many of the projected optical applications of CVD diamond are based on the known intrinsic properties of diamond. The intrinsic absorption mechanisms of diamond are due to interband transitions of electrons across the fundamental electronic energy gap at 5.45 eV and to multiphonon generation in the infrared between 2.5 and 6.5 μm. Generally, absorption due to 2-phonon and 3-phonon generation predominates; higher order phonon generation has significantly lower probability. Absorption due to generation of single phonons is forbidden because of crystal symmetry. Thus, perfect diamond is transparent between 225 nm in the ultraviolet and 2.5 μm in the infrared, and from 6.5 μm in the infrared to zero frequency (dc). Because the absorption process in the infrared between 2.5 and 6.5 μm is of second order and higher, the absorption coefficients are small so that thin diamond films may transmit adequately for many infrared applications in this wavelength range.

CVD diamond is not yet of the quality of the best natural single crystals because of the high density of defects in the material.[28] Several types of defects are present. Lattice defects, such as grain boundaries, twin boundaries, stacking faults, and dislocations, break the crystal symmetry of the diamond lattice resulting in absorption due to generation of single phonons. Thus, absorption by single-phonon generation in CVD diamond has been observed at wavelengths longer than 6.5 μm. Many of the principal features in the diamond phonon spectrum have been identified in these transmission spectra.[28]

Impurities can produce electronic states within the electronic band gap or can lead to local vibrational modes resulting in unwanted absorption. Nitrogen is known to limit significantly the transmission of diamond in the ultraviolet and is responsible for absorption features in the infrared.[29] Hydrogen in combination with carbon produces a C–H stretch absorption feature in the infrared near 2800 cm^{-1} (see References 28, 30–38). Other spectral features due to hydrogen have been observed in bulk diamond and may be present in CVD diamond.[39]

Figure 4.2 shows an infrared absorption spectrum of a CVD diamond polycrystalline specimen prepared by microwave plasma CVD showing the 1-phonon, 2-phonon, and 3-phonon absorption regions and absorption due to a C–H stretch mode.[28]

CVD diamond usually contains nondiamond carbon phases that induce absorption, especially in the visible and ultraviolet. Figure 4.3 shows the absorption coefficient of CVD diamond and of type IIa diamond in the vicinity of the indirect absorption edge near 5.5 eV.[40] The higher absorption coefficient and the absorption tail extending to lower photon energies for the films is believed due to disordered carbon.

The presence of nondiamond phases of carbon is usually detected by Raman spectroscopy. The Raman spectrum of pure diamond consists of a single peak located at 1332 cm^{-1} wavenumber shift. CVD diamond usually exhibits an additional broad peak near 1500 cm^{-1} that is attributed to the nondiamond carbon phases.

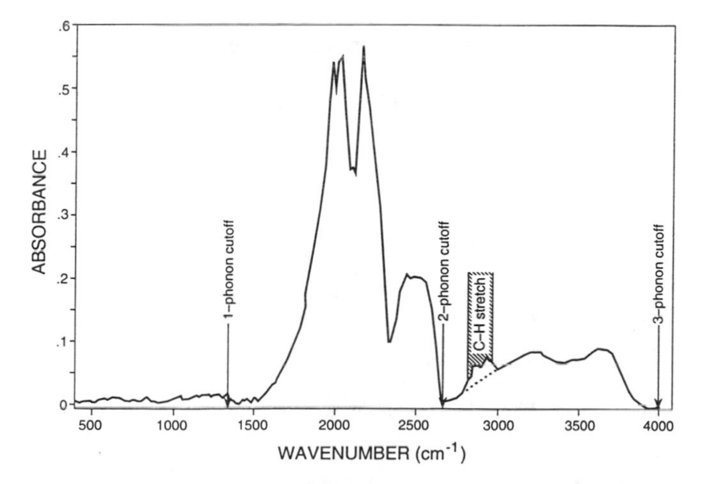

Figure 4.2 **Infrared absorptance spectrum of a CVD diamond specimen prepared by microwave plasma CVD. (Courtesy of C. Klein of the Raytheon Company.[28])**

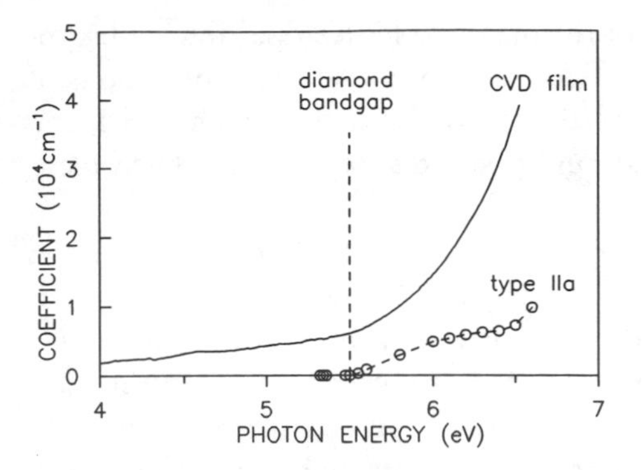

Figure 4.3 Absorption coefficients of a CVD diamond film and a type IIa diamond in the vicinity of the absorption edge.

The relative sizes of the two peaks (after substraction for any luminescence background signal) can be used as a qualitative measure of the diamond quality. Figure 4.4 shows a Raman spectrum typical of CVD diamond.

Recent experiments have shown that diamond can be made highly transmissive in the far infrared. Attenuation of the optical signal was attributed both to free carrier absorption and to optical scatter.[38]

4.4 Defects in CVD Diamond

Defects such as nondiamond phases of carbon, point defects, impurities, and lattice imperfections are an important factor in the optical performance of CVD diamond. Cathodoluminescence imaging and spectroscopy provides a means for examining point defects and impurities in diamond on a microscopic scale.[41] In this method, the specimen is placed in an SEM and the optical radiation emitted by the specimen is collected by a photodetector. The optical signal arises when valence band electrons are excited above the fundamental energy gap of the diamond into the

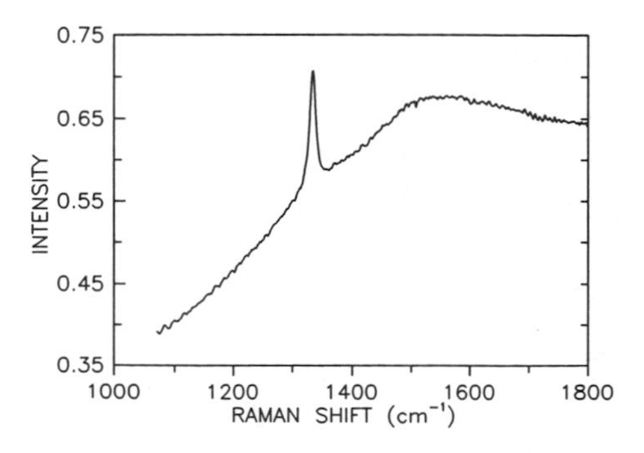

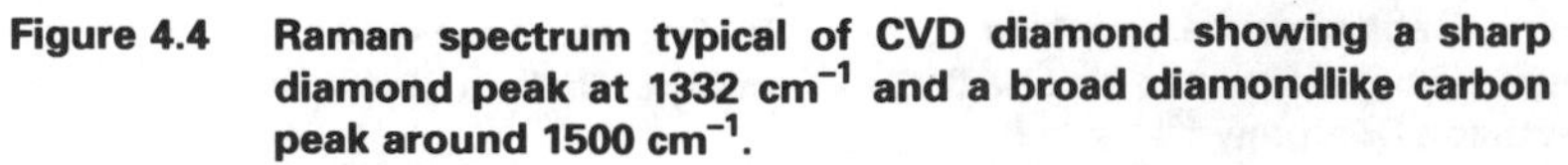

Figure 4.4 Raman spectrum typical of CVD diamond showing a sharp diamond peak at 1332 cm^{-1} and a broad diamondlike carbon peak around 1500 cm^{-1}.

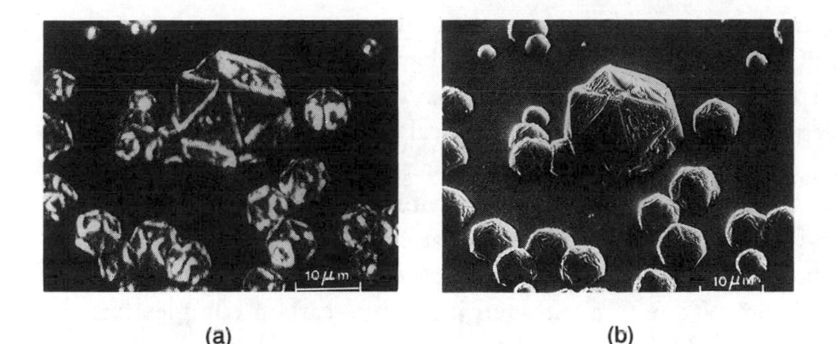

Figure 4.5 **Cathodoluminescence image of (*a*) diamond particles and (*b*) the corresponding SEM image.**

conduction band by the energetic electron beam (>10 kV) of the SEM. The electrons in the conduction band can decay to defect states within the band gap. The electrons may lose energy from these defect states by emitting optical radiation whose spectral features are characteristic of the defect center.

The cathodoluminescence provides an optical image of the specimen as the electron beam scans over the specimen. This image can be compared to the secondary electron image customarily observed with the SEM. Figure 4.5 shows a cathodoluminescence image and a secondary electron SEM image of diamond particles deposited by CVD. The cathodoluminescence image provides information regarding the distribution of luminescent defects in the diamond. However, the interpretation of the image in terms of defect densities is not necessarily straightforward because the intensity of the luminescence from a particular defect species depends not only on the number density of that species but also on the densities of other species that give rise to competing decay processes.

By a spectral analysis of the cathodoluminescence, one can deduce the nature of the defect centers. This identification is based on a large body of work in which the luminescence spectra of many defects have been identified. Figure 4.6 shows the

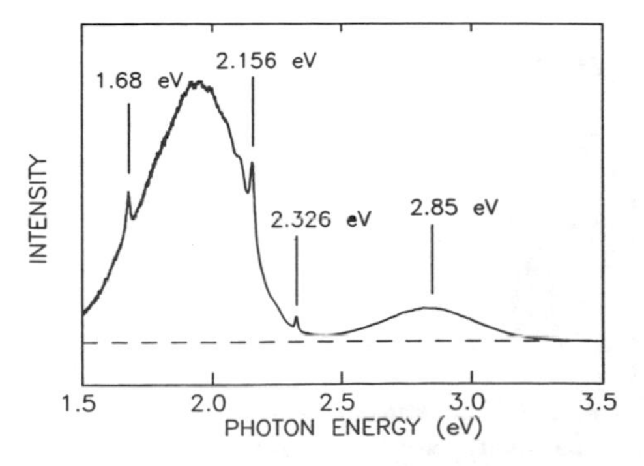

Figure 4.6 **Cathodoluminescence spectrum of CVD diamond film showing principal spectral features.**

cathodoluminescence spectrum of a diamond film having several spectral features often observed in CVD diamond. These features have been associated with particular defects: a sharp line at 1.68 eV believed to be due to a silicon impurity introduced during deposition (this feature is identical to the luminescence line frequently observed at a 5890 cm^{-1} shift in the Raman spectrum using 514.5 nm excitation); a line at 2.156 eV and an associated vibronic band centered near 2 eV, due to a nitrogen-vacancy (N-V) complex; a line at 2.326 eV due to a different N-V complex; and, a broad violet band centered at 2.85 eV, due to a dislocation related defect. A line at 3.188 eV due to a nitrogen interstitial-carbon complex has also been observed in some CVD diamond films.

High-resolution electron microscopy is a powerful tool for examining defects in CVD diamond at the lattice scale.[42] Figure 4.7 shows a high-resolution electron micrograph of CVD diamond with a <110> orientation showing a large number of twinning defects. The film was prepared by microwave plasma technique and thinned for examination in the electron microscope. All of the observed twins retain a crystallographical relationship to each other. All arrows point to misfit twin boundaries, which are described below. The hollow arrows also point along local growth directions.

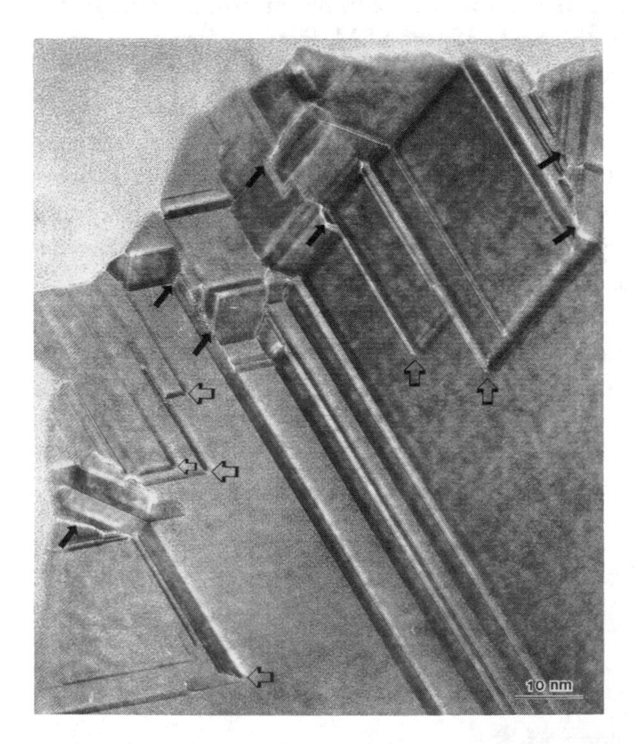

Figure 4.7 **High-resolution electron micrograph of CVD diamond with a <110> orientation. The arrows point to misfit twin boundaries, surfaces where twins grow together noncoherently. The hollow arrows indicate local growth directions.**

A preferred nucleation site for diamond growth is a location at which five twins come together to a point in a structure called a twin quintuplet. Figure 4.8 is an enlarged view of a twin quintuplet and several other twins. A coherent twin boundary occurs when {111} planes of adjacent twins are in coincidence, and this is evident for most of the twin boundaries in the figure.

Because the angle between {111} planes in a crystal twin is 70.5°, a 7.5° angular mismatch occurs at one of the boundaries in a twin quintuplet; this twin boundary is not coherent and is called a misfit boundary. In the figure, the arrow labeled A points to the beginning of a misfit twin boundary that meanders toward the periphery of the crystal. A misfit boundary can terminate within the crystallite or it can extend to the crystallite surface. Because the misfit boundary is the surface at which the crystal growth terminates locally, the local growth direction points along the boundary away from the nucleation site.

4.5 Polishing CVD Diamond

CVD diamond films usually grow with surfaces that are undesirable for most optical application because of a large surface roughness. Smooth diamond films can be made if the nucleation density is high; however, the thickness of such films is limited to several micrometers because the roughness tends to increase with increasing film thickness. However, in a recent paper, Wild et al.[43] have reported that diamond films grown with a <100> texture appear to grow smoother as the film thickness increases. Films of this type have been grown to thicknesses greater than 100 μm. Transmittance measurements indicate that these films show a higher optical transmittance than conventionally grown CVD diamond films.

Methods of polishing CVD diamond films are being developed to produce smooth films. Because CVD diamond is polycrystalline and hard, it is very difficult to

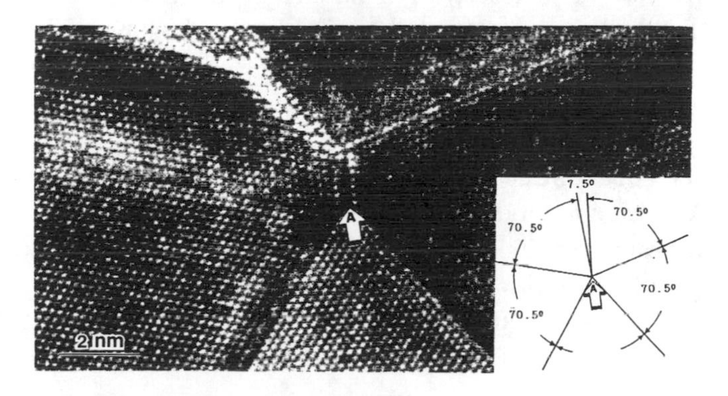

Figure 4.8 **High-resolution electron micrograph showing a twin quintuplet. The arrow is pointing to the start of a misfit boundary. A misfit boundary is a surface at which two twins grow together non-coherently. The arrow also points in a direction of local growth.**

polish; polishing by conventional methods is very slow. Wang et al.[35] have polished CVD diamond films on a cast-iron scaife heated to 350 °C. Six weeks of polishing were required to obtain a mirror-like surface. To increase the polishing rate, they annealed a sample in an atmosphere of 0.01% oxygen in argon at 1000 °C for 4 h; the film surface turned black. In this case, the time for polishing was reduced to one week. Polishing with potassium nitrate also increased the polishing rate; however, the specimen had to be carefully monitored to avoid destruction. Polishing decreased the peak-to-valley surface roughness from 1.2 μm to less than 0.1 μm.

Yoshikawa[44] has pioneered a thermochemical method for polishing diamond at high rates. In his method, a rotating polishing plate of iron (of low carbon content) or nickel is held at an elevated temperature inside an environmental chamber capable of supporting a vacuum. The CVD diamond surface is polished by holding it in contact with the rotating plate. In an atmosphere of hydrogen, iron produced the highest polishing rate and nickel produced nearly as high a polishing rate. No polishing action was observed with molybdenum plates or with cast-iron plates, and no polishing was observed at 700 °C or lower. At 750 °C and above, the polishing rate increased with increasing temperature. At 950 °C, the entire surface was polished after 20 min. The polishing rate also increased with applied pressure; however, excessively high pressures made the polishing process unstable. Increasing the lapping speed also increased the polishing rate. The average roughness, Ra, obtained on a 7-mm-square specimen was 2.7 nm.

Frequently, the diamond surface is too rough for polishing directly. Yoshikawa has planed the surface of the specimen prior to polishing by irradiating the specimen with a Q-switched Nd-doped yttrium aluminum garnet (Nd:YAG) laser in one atmosphere of oxygen. A peak-to-valley roughness of 3 μm could be obtained by this process. Several authors have used variations of Yoshikawa's method to polish CVD diamond.[45, 46]

Protrusions that sometimes grow on the diamond surface must be removed prior to polishing. Harker et al.[45] have used reactive plasma etching to remove such protrusions. In order to etch only the protrusions and not the surrounding material, they applied a nonreactive gold coating to the entire surface. The protrusions were then exposed for reactive plasma etching with oxygen.

Another method of polishing CVD diamond with an ion beam has been developed.[47] The rough surface of a diamond specimen is spin coated with a mixture of photoresist and a Ti–silica emulsion to produce a plane surface. The surface is then etched with an oxygen ion beam; the angle of incidence is chosen to match the etching rate of the coating with the etching rate of diamond. The average root-mean-squared roughness on a 5-cm-diameter film was 4.9 nm.

4.6 X-ray Window

Because of its low atomic number, carbon is highly transparent to X-ray radiation. The excellent mechanical properties of diamond make it useful as an ultrathin

X-ray window for an energy-dispersive X-ray fluorescence detector.[4] This has been the first practical use of CVD diamond for tranmitting electromagnetic radiation. In this application, diamond is replacing beryllium, which must be made considerably thicker to support a vacuum. The optical transparency and mechanical stability of diamond also make CVD diamond a possible membrane material for X-ray masks. X-ray masks are used in the X-ray lithography of integrated circuits.[5]

4.7 Summary

The superior properties of diamond make it a candidate for a number of optical applications. The most immediate application is the diamond X-ray window. The quality of CVD diamond for other applications, such as infrared transmissive elements and coatings, is continually improving but is not yet sufficient. Surface roughness is a major impediment to optical applications. The ability to deposit smooth surfaces would make diamond considerably more attractive as an optical material. Polishing is being pursued as a means for making smooth CVD diamond surfaces. Diamond also has promise as a blue luminescent or laser material. Identifying and controlling the relevant luminescent defect centers will be needed to improve the quantum efficiencies of such devices.

Acknowledgments

The work in this chapter was supported in part by the Office of Naval Research. We thank J. L. Hutchison of Oxford University for his assistance in obtaining the high resolution electron micrographs.

References

1 V. P. Varnin, B. V. Deryagin, D. V. Fedoseev, I. G. Teremetskaya, and A. N. Khodan. *Sov. Phys. Crystallogr.* **22**, 513–515, 1977.

2 B. V. Spitsyn, L. L. Bouilov, and B. V. Derjaguin. *J. Cryst. Growth.* **52**, 219–226, 1981.

3 S. Matsumoto, Y. Sato, M. Kamo, and N. Setaka. *Jpn. J. Appl. Phys.* **21**, L183, 1982.

4 M. G. Peters, J. L. Knowles, M. Breen, and J. McCarthy. In "Diamond Optics II." *Proceedings*, No. 1146. (A. Feldman and S. Holly, Eds.) SPIE, 1990, pp. 217–224.

5 H. Windischmann and G. F. Epps. *J. Appl. Phys.* **68**, 5665–5673, 1990.6 Y. Taniguchi, K. Hirabayashi, K. Ikoma, N. Iwasaki, K. Kurihara, and M. Matsushima. *Jpn. J. Appl. Phys.* **28**, L1848–L1850, 1989.

7 S. C. Rand and L. G. DeShaser. *Opt. Lett.* **10**, 481–483, 1985.

8 C. B. Beetz, Jr., B. A. Lincoln, and D. R. Winn. In "Diamond Optics III." *Proceedings*, No. 1325. (A. Feldman and S. Holly, Eds.) SPIE, 1990, 240–252.

9 W. D. Partlow, R. E. Witkowski, and J. P. McHugh. In *Applications of Diamond Films and Related Materials.* (Y. Tzeng, M. Yoshikawa, M. Murakawa, and A. Feldman, Eds.) Elsevier, Amsterdam, 1991, pp. 163–168.

10 M. Kamo, Y. Sato, S. Matsumoto, and N. Setaka. *J. Cryst. Growth.* **62**, 642–644, 1983.

11 K. Suzuki, J. Yasuda, and T. Inuzuka. *Appl. Phys. Lett.* **50**, 728–729, 1987.

12 S. Matsumoto. *Proceedings.* 7th International Symposium Plasma Chem., 1985, pp. 79–84.

13 S. Matsumoto, M. Hino, and T. Kobayashi. *Appl. Phys. Lett.* **51**, 737–739, 1987.

14 A. B. Harker. Presentation at the SPIE Diamond Optics IV Conference, San Diego, July 1991.

15 Y. Hirose and N. Kondo. Program and Book of Abstracts. Japan Applied Physics 1988 Spring Meeting, 29 Mar. 1988, p. 434.

16 L. M. Hanssen, W. A. Carrington, J. E. Butler, and K. A. Snail. *Materials Letters.* **7**, 289–292, 1991.

17 E. N. Farabaugh, A. Feldman, and L. H. Robins. In *Applications of Diamond Films and Related Materials.* (Y. Tzeng, M. Yoshikawa, M. Murakawa, and A. Feldman, Eds.) Elsevier, Amsterdam, 1991, pp. 483–488.

18 Y. Hirose and Y. Terasawa. *Jpn. J. Appl. Phy. Part 2.* **25**, L519, 1986.

19 C.-P. Chiang, D. L. Flamm, D. E. Ibbotson, and J. A. Mucha. *J. Appl. Phys.* **63**, 1744–1748, 1988.

20 D. E. Meyer, R. O. Dillon, and J. A. Woollam. In *Diamond and Diamond-Like Films.* (J. P. Dismukes, Ed.) The Electrochemical Society, Pennington, NJ, 1989, pp. 494–499.

21 T. P. Ong and R. P. H. Chang. *Appl. Phys. Lett.* **55**, 2063–2065, 1989.

22 J. W. Glesener, A. A. Morrish, and K. A. Snail. In *Applications of Diamond Films and Related Materials.* (Y. Tzeng, M. Yoshikawa, M. Murakawa, and A. Feldman, Eds.) Elsevier, Amsterdam, 1991, pp. 347–351.

23 K. A. Snail and L. M. Hanssen. *J. Crystal Growth.* **112**, 651–659, 1991.

24 M. W. Geis, H. I. Smith, A. Argoit, J. Angus, G.-H. M. Ma, J. T. Glass, J. Butler, C. J. Robinson, and R. Pryor. *Appl. Phys. Lett.* **58**, 2485–2487, 1991.

25 J. F. Prins and H. L. Gaigher. In "New Diamond Science and Technology." *Proceedings.* The Second International Conference. (R. Messier, J. T. Glass, J. E. Butler, and R. Roy, Eds.) Materials Research Society, Pittsburgh, 1991, pp. 561–566.

26 J. Narayan, V. P. Godbole, and C. W. White. *Science.* **252**, 416–418, 1991.

27 D. E. Patterson, B. J. Bai, C. J. Chu, R. H. Hauge, and J. L. Margrave. In "New Diamond Science and Technology." *Proceedings.* The Second International Conference. (R. Messier, J. T. Glass, J. E. Butler, and R. Roy, Eds.) Materials Research Society, Pittsburgh, 1991, pp. 433–438.

28 C. Klein, T. Hartnet, R. Miller, and C. Robinson. In "Diamond Materials." *Proceedings.* The Second Intern'l Symp. on Diamond Materials, 1991. (A. J. Purdes, J. C. Angus, D. F. Davis, B. M. Meyerson, K. E. Spear, and M. Yoder, Eds.) The Electrochemical Society, Pennington, NJ, 1991, pp. 435–442.

29 S. Musikant. *Optical Materials.* Marcel Dekker, New York and Basel, 1985, pp. 113–115.

30 C. E. Johnson and W. A. Weimer. In "Diamond Optics II." *Proceedings,* No. 1146. (A. Feldman and S. Holly, Eds.) SPIE, 1990, pp. 188–191.

31 T. Feng. In "Diamond Optics II." *Proceedings,* No. 1146. (A. Feldman and S. Holly, Eds.) SPIE, 1990, pp. 159–165.

32 K. A. Snail, L. M. Hanssen, A. A. Morrish, and W. A. Carrington. In "Diamond Optics II." *Proceedings,* No. 1146. (A. Feldman and S. Holly, Eds.) SPIE, 1990, pp. 144–151.

33 Y. Cong, R. W. Collins, G. F. Epps, and H. Windischmann. *Appl. Phys. Lett.* **58**, 819–821, 1991.

34 M. A. Akerman, J. R. McNeely, and R. E. Clausing. In "Diamond Optics III." *Proceedings,* No. 1325. (A. Feldman and S. Holly, Eds.) SPIE, 1990, pp. 178–186.

35 X. H. Wang, L. Pilione, W. Zhu, W. Yarbrough, W. Drawl, and R. Messier. In "Diamond Optics III." *Proceedings,* No. 1325. (A. Feldman and S. Holly, Eds.) SPIE, 1990, pp. 160–167.

36 X. X. Bi, P. C. Eklund, J. G. Zhang, A. M. Rao, T. A. Perry, and C. P. Beetz, Jr. *J. Mater. Res.* **5**, 811–817, 1990.

37 X. X. Bi, P. C. Eklund, J. G. Zhang, A. M. Rao, T. A. Perry, and C. P. Beetz, Jr. In "Diamond Optics II." *Proceedings,* No. 1146. (A. Feldman and S. Holly, Eds.) SPIE, 1990, pp. 192–200.

38 A. J. Gatesman, R. H. Giles, J. Waldman, L. P. Bourget, and R. Post. In "Diamond Optics III." *Proceedings,* No. 1325. (A. Feldman and S. Holly, Eds.) SPIE, 1990, pp. 170–177.

39 E. Fritsch and D. V. G. Scarratt. In "Diamond Optics II." *Proceedings,* No. 1146. (A. Feldman and S. Holly, Eds.) SPIE, 1990, pp. 201–206.

40 L. H. Robins, E. N. Farabaugh, and A. Feldman. In "Diamond Optics IV." *Proceedings,* No. 1534. (A. Feldman and S. Holly, Eds.) SPIE, 1991, pp. 105–116.

41 L. H. Robins, L. P. Cook, E. N. Farabaugh, and A. Feldman. *Phys. Rev. B.* **39**, 13367–13377, 1989-II.

42 D. Shechtman, E. N. Farabaugh, L. H. Robins, A. Feldman, and J. L. Hutchison. In "Diamond Optics IV." *Proceedings*, No. 1534. (A. Feldman and S. Holly, Eds.) SPIE, 1991, pp. 26–43.

43 C. Wild, W. Muller-Sebert, T. Eckermann, and P. Koidl. In *Applications of Diamond Films and Related Materials.* (Y. Tzeng, M. Yoshikawa, M. Murakawa, and A. Feldman, Eds.) Elsevier, Amsterdam, 1991, pp. 197–205.

44 M. Yoshikawa. In "Diamond Optics III." *Proceedings*, No. 1325. (A. Feldman and S. Holly, Eds.) SPIE, 1990, pp. 210–221.

45 A. B. Harker, J. Flintoff, J. F. and DeNatale. In "Diamond Optics III." *Proceedings*, No. 1325. (A. Feldman and S. Holly, Eds.) SPIE, 1990, pp. 222–229.

46 T. P. Thorpe, A. A. Morrish, L. M. Hanssen, J. E. Butler, and K. A. Snail. In "Diamond Optics III." *Proceedings*, No. 1325. (A. Feldman and S. Holly, Eds.) SPIE, 1990, pp. 230–237.

47 B. G. Bovard, T. Zhao, and H. A. Macleod. In "Diamond Optics III." *Proceedings*, No. 1325. (A. Feldman and S. Holly, Eds.) SPIE, 1990, pp. 216–222.

Part II

Stability and Modification of Film and Surface Optical Properties

5

Multilayer Optical Coatings

PETER M. MARTIN

Contents

5.1 Introduction

Multilayer optical coatings are stacked layers of thin-film optical materials designed to perform a specific function in a given spectral range. Figure 5.1 shows a schematic cross-sectional view of the layer structure of a multilayer coating. The layers of the coating usually are composed of two or three different materials stacked in a prescribed sequence. The coatings can be composed of dielectric, semiconductor, and metal layers, and each material may have a different microstructure. Applications include all-dielectric and dielectric-enhanced metal high reflectors, antireflection coatings, beam splitters, polarizers, edge filters, and rugate filters.

This chapter describes how composition and microstructure determine the optical performance of the individual layers and multilayer optical coatings, the common techniques used to determine composition and microstructure, and when to use them.

The optical performance of multilayer optical coatings results from interference effects due to fresnel reflections at each layer interface; it is affected by optical absorption in each layer and at each interface and optical scattering due to coating morphology. The optical constants of the individual layer materials, the refractive index (n), the extinction coefficient (k), and the absorption coefficient (β), and the

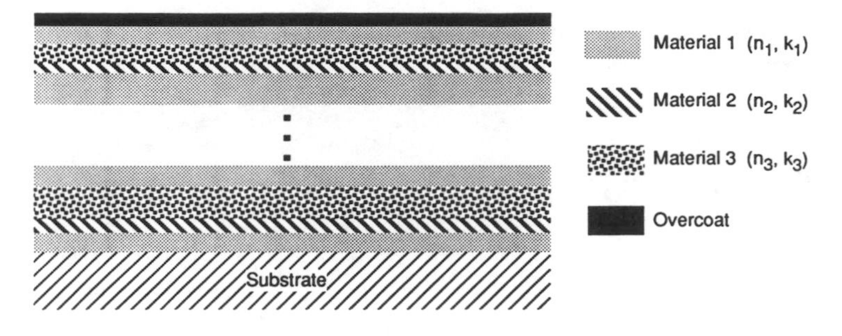

Figure 5.1 Schematic cross section of multilayer optical coating.

individual layer thicknesses determine the optical performance and functional spectral range of the coating. Layers with high- and low-refractive index are alternated to create the fresnel reflections and the resulting interface effects. The quarterwave stack is the simplest multilayer optical coating to fabricate and evaluate. In this case, all high- and low-index layers have the same optical thickness. The non-quarterwave stack, in which no layer has the same optical thickness, is more common and makes it more difficult to determine composition and evaluate optical performance. Interfaces between layers can be discrete or have their composition graded between that of the adjacent layers.

The composition of each thin-film layer determines n, k, and β; small variations in layer composition can significantly change these parameters and the optical performance of a multilayer coating. Small changes in n and k can result from compositional nonuniformities and inhomogeneities in each layer. To a lesser degree, the microstructure of each layer affects n, k, β, and the optical performance of the coating as a whole. To obtain the correct, or design, optical performance, researchers must know and precisely control the composition of each layer. For example, n of hydrogenated amorphous silicon (Si:H) coatings, used in near-infrared (NIR) and infrared (IR) multilayers, depends strongly on composition and decreases from 3.8 to 2.5 with an increase in H content from 0 to 35 at. %.[1] Optical absorption also decreases, and k decreases from 0.30 to less than 0.0002. The optical constants of most oxide and nitride optical coatings, and their dispersion with wavelength, also depend strongly on composition.

When possible, it is preferable and less complicated to determine the composition, microstructure, and optical constants of the individual layers of a multilayer coating. This is because interference effects in multilayer coatings enhance absorption and the associated changes in optical performance resulting from compositional and microstructural variations (and the resulting changes in n, k, and β). Unless the exact coating design and layer thicknesses are known, exact determination of composition is difficult. The measurements of the composition of layers near the bottom of the stack may be masked by upper layers.

The most important and often-used techniques for determining composition of single-layer and multilayer optical coatings are optical spectroscopy (absorption, transmittance, and reflectance), surface (Auger electron spectroscopy [AES], secondary ion mass spectroscopy [SIMS], and nuclear reaction analyses), and chromatography (gas evolution). Each has its advantages and limitations, and not all will provide enough information to determine uniquely the composition of a coating. It is often desirable to use several techniques to obtain unambiguous results.

The microstructure of thin-film optical coatings can range from amorphous to polycrystalline. Sometimes single-crystalline epitaxial coatings are required. Each type of microstructure has its advantages and disadvantages.

Coating microstructure directly affects optical scattering and coating stability, and physisorbed impurities can alter n and k and the measured optical response. The presence of voids ($n = 1$, $k = 0$) and mixed-crystalline phases change n and k by their volume fraction (to first order). The surface morphology of amorphous coatings is generally smoother than that of polycrystalline coatings, resulting in less optical scattering. Because there are no low-density grain boundaries, amorphous coatings are generally more environmentally stable. Polycrystalline coatings may have two or more crystalline phases present and are usually harder than amorphous coatings.

Choice of characterization technique will also depend on whether or not the coating can be physically or compositionally modified, damaged, destroyed, or altered when placed in vacuum. Spectroscopic and nuclear reaction techniques are nondestructive, while surface techniques require milling of the coating, and chromatography breaks the coating apart. Coating microstructure is determined by optical microscopy, transmission electron microscopy (TEM), scanning transmission electron microscopy (STEM), scanning electron microscopy (SEM), electron diffraction, X-ray diffraction (XRD), Raman spectroscopy, and, less importantly, by optical spectroscopic techniques. Choice of technique will depend on the information desired and is discussed in detail in the text. TEM and STEM analyses require complicated sample preparation techniques, and SEM requires that the sample be conductive, or the application of a thin-conductive overcoat may be necessary. XRD analysis is usually nondestructive.

Evaluation of the composition and microstructure of single-layer optical coatings is discussed first, followed by the same treatment for the more complex case of multilayer optical coatings. In the final section, the environmental stability of multilayer optical coatings is addressed.

5.2 Single-Layer Optical Coatings

Although a multilayer optical coating is composed of alternating high- and low-refractive-index materials, the optical performance (or properties) and composition of an individual layer are difficult to determine due to interference effects which occur between all the layers, uncertain and unequal layer thicknesses, and the masking of bottom layers by optical absorption in the top layers. Analysts can best understand techniques used to determine the composition and optical properties of multilayer coatings by first examining the techniques used to evaluate the constituent single-layer coatings, and then applying them to multilayer coatings. Many of the techniques used to determine microstructure in single-layer coatings can be used directly for multilayer coatings. In most cases, it is preferable to evaluate multilayer coatings because the microstructure may not be the same as that of a separate single layer.

Optical Constants

The composition of an optical coating is the primary factor determining the optical constants *n* and *k* and their dispersion relations. The mole fraction of each constituent also determines the optical band gap, the absorption coefficient, and sometimes the bonding configuration and microstructure. The next two examples demonstrate the need to determine and control the coating composition accurately to obtain the desired optical properties. Figure 5.2 shows the dependence of *n* at the 0.63-μm wavelength of sputtered oxynitride coatings on the mole fractions of $(SiO_2)_x$ and $(Si_3N_4)_{1-x}$ measured by Auger electron spectroscopy.[2] The *n* is very sensitive to the mole fraction of SiO_2 and decreases from 1.98 for $x = 0$ (all Si_3N_4) to 1.46 for $x = 1$ (all SiO_2). Note that either constituent may act as an impurity and cause *n* to vary from the desired value.

The *k* of most dielectric and compound semiconductor materials is very sensitive to stoichiometry and composition. There is always a composition which has the lowest optical absorption ($k = \beta\lambda/4\pi$). Figure 5.3 shows the dependence of the dispersion of *k* of hydrogenated amorphous Si (Si:H) with H content. In this case, the H content was measured by IR spectroscopic techniques (see the optical spectroscopy section). For wavelengths below 2.5 μm, the coating optical absorption and *k* decrease with increased H content. Because the lowest possible absorption is usually

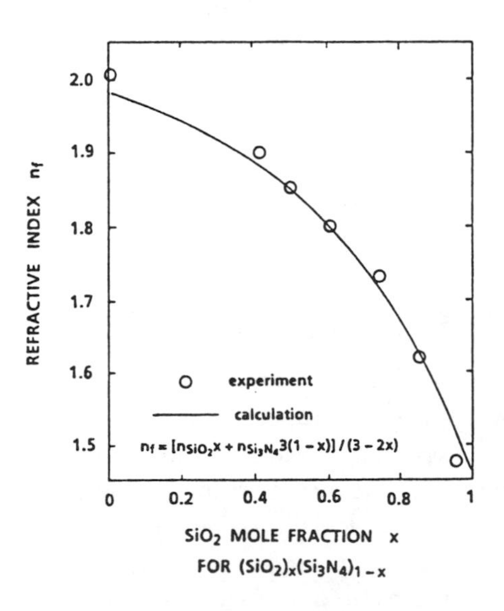

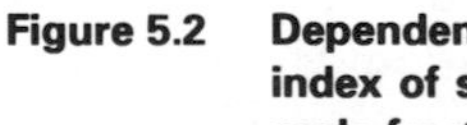

Figure 5.2 Dependence of the refractive index of silicon oxynitride on mole fraction of oxygen.[2]

required to meet optical performance goals, Si:H coatings with the highest possible H content should be used.

Coating microstructure can also affect n and k, but usually to a lesser degree than composition. Porosity, and the resultant physisorbed impurities, such as water, change both n and k. The refractive index of the coating is usually decreased by the volume fraction of the physisorbed water and voids.[3]

Impurities cause additional, and usually unwanted, absorptions in many wavelength regions. In a polycrystalline coating, to first order, both n and k are also changed by the volume fraction of the crystalline phases present. Phase composition has been shown to modify n of polycrystalline TiO_2 coatings, increasing approximately linearly from the n of pure anatase to that of pure rutile.[4] Grains and grain boundaries usually reduce n and increase k compared with bulk crystalline materials.

Composition Measurement Techniques

No single analytical technique is ideally suited for all composition measurements. The choice of measurement technique to evaluate coating composition and associated chemical bonding depends on the accuracy required, the type of information needed (depth profile, etc.), whether or not the coating can be damaged or chemically modified, and the availability of the technique. The four general types of techniques used to determine composition are

- optical spectroscopic techniques
- photo-induced phenomena
- surface reaction techniques
- nuclear reaction techniques.

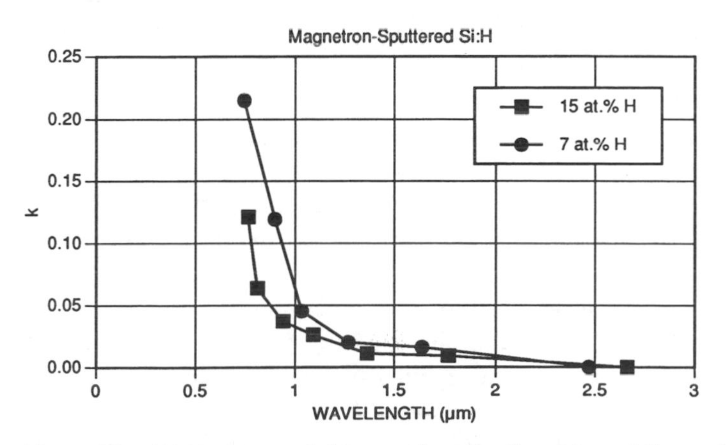

Figure 5.3 Dependence of the wavelength dispersion of the extinction coefficient of Si:H alloys for H contents of 7 and 15 at. %.

The most widely used techniques in each category and their application are described in the following subsections.

Optical spectroscopic techniques Optical spectroscopic techniques used to determine chemical composition and bonding, particularly those using absorption of IR radiation, have reached a state of maturity, are well understood, and are extensively used. This technique is often used for hydride, carbide, nitride, sulfide, fluoride, and oxide coatings. Oxides of Si, Ge, Ti, Mg, In, Ta, V, Fe, Zn, Y, Al, and Cr all display IR spectral bands. If available, this is an inexpensive technique to evaluate composition. The IR transmission and absorption spectra of a specific bulk composition or functional group of elements, if they are IR transparent, are usually unique enough to identify, or "fingerprint," that compound. A multitude of handbooks, publications, and references which identify IR–spectral-absorption bands for a wide range of compounds are readily available (see Reference 5, for example). Although it is not as straightforward, this technique has been adapted from its primary use in analyzing composition and molecular structure of bulk chemical liquids for use with thin-film optical coatings. The analysis of optical coatings is usually more straightforward because there are fewer constituents and the identities of these constituents and their possible bonding configurations are usually known.

The composition determined by this technique is indicative of the entire volume average of each constituent of the coating. It is not possible to profile each composition through the thickness of the coating, which is a major weakness of this technique for multilayer applications. Raman spectroscopy, the complementary optical-analysis technique used to evaluate chemical bonding in the coating, is described in a later section.

It is beyond the scope of this book to treat the processes and mechanisms of molecular optical absorption theoretically. Detailed treatments are available which fully describe the absorption process, the origin of IR absorption bands, the number of absorption bands, the position of absorption bands, and quantitative analysis of chemical composition (see References 6 and 7, for example). Numerous treatments for thin films with specific compositions appear in the literature.

A dual-beam IR spectrophotometer is the standard instrument used to measure the IR transmission and absorption spectra of an optical coating. Typical measurement wavelengths range from 2 to 200 μm. For this technique to be useful, the optical coating must be transparent, or not strongly absorbing, over all or part of this wavelength range.

To obtain maximum information, analysts must determine the intensity of the absorption, the integrated area under the absorption band, and the wavelength position of the absorption either digitally or graphically from the measured spectra. The identity of the vibrational mode is determined from the wavelength or wavelength band of the absorption peak or peaks, and the mole fraction of each constituent is determined from the integrated absorption intensity under each peak.

The same molecular-bond type may have several IR absorptions, depending on the types of vibrational modes stimulated (symmetric and asymmetric bond stretching, bond wagging, bond twisting and rocking, or symmetric and asymmetric bond bending). Also, the vibrational modes of similar bond types may overlap and must be deconvoluted.

The literature should first be consulted to obtain the position of the absorption bands and calibration data for composition measurements of the optical coating being evaluated. If data or literature is not available for the coating, the analyst should begin by referencing the absorption spectrum of the bulk material to get an idea of the position of the absorption bands. The exact wavelength of the absorption in the coating may differ from that of the bulk due to differences in and lack of crystalline order and differences in the local environment of the constituents. Also, the calibration between absorption intensity and mole fraction of the coating constituents depends on oscillator strength (dipole moment) of the bonds and will differ from bulk. Unless the integrated area under the absorption band is calibrated against a known mole fraction, only relative concentrations can be obtained.

Quantitative analytical techniques to determine composition of thin films appear in many handbooks and in the literature.[5, 8] To demonstrate compositional analysis by this technique, Figure 5.4*a* shows an IR-transmittance spectrum of a Si:H coating on a Ge substrate. Note that, in addition to the interference pattern due to fresnel reflections at each side of the coating, absorption bands occur near 2090, 2000, 897, 847, and 660 cm^{-1}. These are the Si–H bond stretching, bending, and wagging vibrational modes, respectively, which have been assigned previously in the literature.[9] It is generally assumed that the observed IR bands are associated with allowed vibrational transitions intrinsic to the material. Figure 5.4*b* shows the absorption band centered at 2000 cm^{-1}, with a shoulder centered at 2090 cm^{-1}. The absorption band centered at 2000 cm^{-1} is attributed to the Si–H bond configuration, and the absorption band at 2090 cm^{-1} is attributed to the $Si–H_2$ bond type. A careful gaussian analysis is needed to separate the broad absorption band into the two individual components. Once the oscillator strength is calibrated, H concentration is obtained directly from the integrated area associated with either the stretching or bending mode. The oscillator strength must be known for the absorption band being used. The H content of the coating shown is 33 at. %. This measurement was calibrated using SIMS and AES. Other calibration measurements are discussed in the following sections.

IR spectroscopic techniques also provide detailed information about the molecular bonding configuration or configurations in the coating. The center wavelength of the absorption is representative of an individual bond configuration. Let us continue with the Si:H example. The Si–H bond type displays absorptions at 2000 and 645 cm^{-1}, and the $Si–H_2$ bond type displays absorptions at 2090, 880, and 650 cm^{-1}. If these wavelengths were not well known from literature, the analyst should first estimate the IR vibrational absorption wavelengths by consulting the

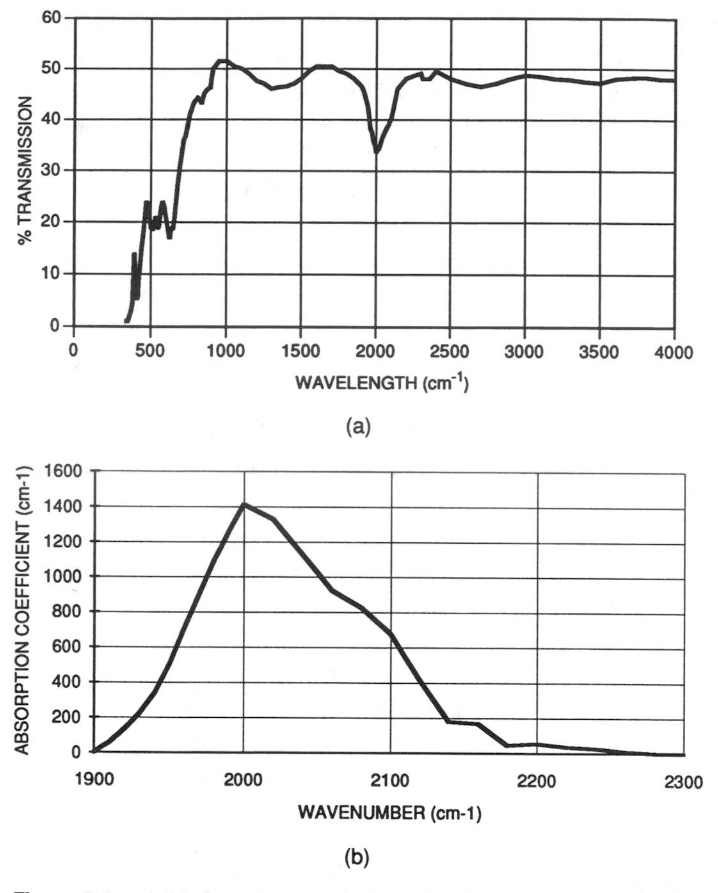

Figure 5.4 **(*a*) Infrared transmission of a Si:H coating on a Ge substrate. (*b*) Absorption spectrum of a SiH coating, calculated from the transmission spectrum in graph *a*.**

handbook values for silane (SiH_4) and its derivatives. Other examples of optical coatings with multiple bond configurations are diamond-like carbon (DLC),[9] Si_3N_4,[10] GeC,[11] SiC:H,[12] and GaAs:H.

This technique is useful for detecting many impurities in a coating and determining their concentration. Impurities such as water, H, F, O, and N can readily be detected from their IR-absorption bands.[5] Total internal reflection (TIR) absorption spectroscopy is more sensitive than conventional techniques and is often used to detect impurities optically. Monolayer films adsorbed onto metal surfaces have been detected. Here a TIR stage is mounted in the spectrophotometer, and the spectrophotometer beam is directed into the coating at incidence angles greater than the critical angle, creating a total internal reflection. The coating absorbs energy via the interaction with an evanescent wave. Multiple reflections increase the path length of the beam through the coating and enhance the absorption in the coating. The absorption bands are treated in the same manner as with single-pass techniques.

Laser Raman spectroscopy Raman spectroscopy, the complementary optical technique to IR spectroscopy, is used to evaluate molecular structure of thin-film optical coatings. The features of Raman spectra can be used directly to determine molecular-bonding configuration and microstructure.[13] Raman spectroscopy is also a strong tool used in determining crystalline-phase composition and can be used for in situ molecular-structure measurements. Virtually all recent Raman spectroscopic measurements employ high-intensity laser-light sources. Inelastically scattered radiation is imaged into a grating spectrometer and detected, producing the Raman spectrum. Micrometer-size spatial-mapping resolution has been demonstrated.

The Raman spectrum of a coating consists of high-intensity bands, or maxima, and low-intensity regions. The intensity maxima, or peaks, have a particular bandwidth and center frequency. Similar to IR vibrational spectra, the center frequency of a maximum correlates with the frequency shift of the inelastically scattered light and constitutes the Raman spectrum. It is uniquely associated with a particular molecular-vibrational mode in the coating. The frequency position of the maximum depends on both the molecular-bond composition and the local-bond arrangement. The Raman spectrum can be used to determine the crystalline phases present in the coating and their relative concentrations.

With the increased use of multichannel diode-array and charge coupled device (CCD) detectors, the Raman spectrum of a coating can be measured in a matter of seconds. The bonding and microstructural information are obtained from the peak frequencies, bandwidths at half maximum, peak asymmetries, and peak intensities of the Raman spectrum. The Raman intensity spectrum of a thin-film coating may be reduced by as much as a factor of 100 compared to a bulk-crystalline sample. Unless signal enhancement techniques are used, coating thicknesses near 1000 nm (1μm) are needed to obtain spectra with well-defined features. Figure 5.5 shows the Raman spectra for TiO_2 coatings with thicknesses of 143, 427, and 831 nm, and bulk anatase powder.[14] The spectra were excited by the 514.5-nm line of an Ar-ion laser. It should be noted that bulk single-phase samples are used to calibrate the Raman technique for use with thin-film optical coatings.

Salient features of all the spectra are (1) the peak frequency, (2) the bandwidths of the peaks, and (3) the intensities of the peaks. Crystalline phases are identified from the peak frequency or frequencies. The spectra of the two thinner coatings are broad and featureless and primarily show features of the fused silica substrate. The spectrum of the 831-nm-thick coating shows broad peaks at 143, 200, 395, 514, and 635 cm^{-1}, indicative of the anatase crystalline phase of the bulk powder. Peak frequencies can be determined to ± 0.05 cm^{-1}. The presence of only anatase peaks indicates that only this phase is present in the coating. The strongest peaks for the rutile crystalline phase are at 440 and 607 cm^{-1}. Note that the Raman features associated with the coating are broader than those of the single crystal and that the

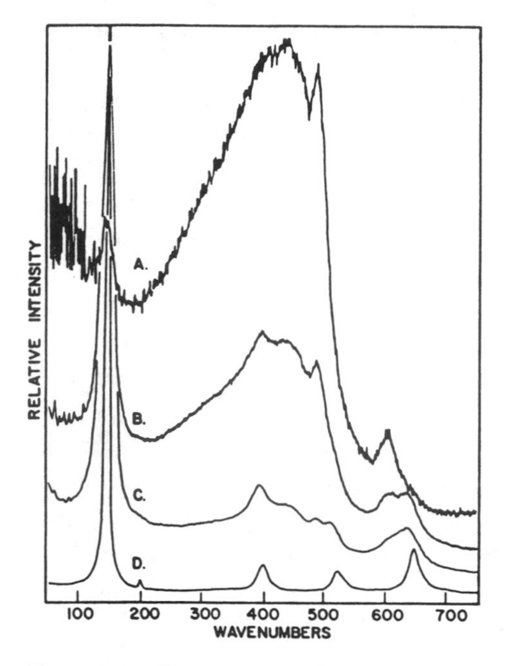

Figure 5.5 **Raman spectra for three single-layer TiO_2 coatings on fused silica substrates compared with the spectrum of bulk anatase TiO_2.**

peaks near 395 and 514 cm^{-1} are convoluted with the fused silica background. The broadening is due to the small grains and crystalline disorder. The bandwidth is inversely proportional to the grain size and will usually increase for coatings over those seen for bulk samples. The peak intensities determine the relative phase composition. In this case, only the anatase phase is present. If both phases were present, the peaks would have to be separated and their intensities determined to provide information about relative phase composition. Asymmetry of the peaks in a Raman spectrum provides information about bond deformation due to internal mechanical and thermal stresses in the coatings.

Raman spectroscopy is frequently used in situ to determine the phase composition or change in phase composition of a coating during deposition, heat treatment, exposure to high-power-laser radiation, and many other processes which can alter phase composition.[13]

Surface analytical techniques Compositional analysis by surface and nuclear analytical techniques involves bombardment of the thin-film coating by electromagnetic radiation (usually X-rays), electrons, or nuclear particles, and detection of the emission of electrons, electromagnetic radiation, or nuclear particles resulting from nuclear, electronic, and electromagnetic interactions in the coating. The surface techniques described in this section are AES, SIMS, X-ray photoelectron spectroscopy (XPS or ESCA), nuclear reaction analysis (NRA), and Rutherford backscattering spectroscopy (RBS). Excellent reviews of all these techniques can be found in the literature.[15–19]

Choice of technique will depend on the particular type of compositional information needed, the elemental sensitivity of each technique, and whether or not destructive analysis is acceptable. AES and RBS are the two most widely used techniques. Several techniques, however, may be used to provide complementary information. Each technique is most sensitive to a specific range of elemental masses. Those chosen should be most sensitive to the expected chemical composition of the coating being evaluated and to the possible impurities. The techniques discussed in this section are particularly useful in evaluating the following properties of common oxide, nitride, carbide, and hydrogen-containing optical materials:

- composition
- composition depth profiles
- interface composition profiles
- impurity profiles
- stoichiometry resulting from deposition conditions
- homogeneity and uniformity of composition over the coating surface
- diffusion, oxidation, and sulfidation.

Table 5.1 gives a comparison of the important characteristics of these surface analytical techniques.[20] Many of these techniques are also used for in situ compositional analysis during deposition and other analyses during which composition is modified. It should also be noted that, except for a few special cases, these techniques yield little information about phase composition or microstructure.

Auger electron spectroscopy (AES) AES is one of the most documented, well-calibrated, widely-used, and available surface analysis techniques. Because it is representative of many surface analytical techniques and the measurement results are similar, it will be presented in more detail than the other techniques. Auger spectrometers are accessible at most industrial and research facilities and universities. A main advantage of AES is that it is sensitive to low-mass impurities, such as carbon and oxygen, which are commonly found at interfaces and surfaces of optical coatings. This technique is sensitive enough to detect adsorption of 0.5 monolayers of oxygen atoms. However, the sensitivity of AES to H and He is poor. The coating must be placed in a vacuum during analysis, which can be time-consuming and cumbersome. Spot sizes down to 30 nm can be sampled. In most cases, the electron beam can be scanned across the coating, allowing detailed and localized compositional analysis of the entire coating. Compositional depth profiles can be produced by sampling while ion milling through the coating. Because this technique bombards the coating with electrons, charging of insulator coatings can occur which decreases measurement sensitivity. Also, the electron beam can damage the coating.

Incident Beam	X-Rays[a]	Electrons[a]	Optical Photons[a]	Charged Particles[a]
X-rays	X-ray diffraction and fluorescence	electron spectroscopy for chemical analysis (ESCA)	—	—
Electrons	electron microprobe, energy-dispersive analysis	TEM, SEM, STEM, LEED, HEED, energy-loss measurements	cathodoluminescence	—
Optical photons	—	photoemission	optical microscopy	—
Charged particles	ion-induced X-ray analysis	ion-induced secondary electron emission	ion-induced photon analysis	Rutherford backscattering, SIMS, nuclear reaction analysis

[a] Scattered or emitted beam for analysis

Table 5.1 Comparison of important features of surface analytical techniques.[16]

Sensitivity may also be decreased because the sputtered, or ion-milled surface, may not be representative of the actual composition due to preferential sputtering. Analysis of the Auger spectra can be tricky, and requires a trained operator.

During the Auger process, an electron is emitted by the decay process when an inner-shell vacancy is filled by an outer-shell electron. The vacancy is caused by the bombardment of a photon, electron, or ion. Energy in the form of X-rays or an ejected outer-shell electron is released when the outer-shell electron decays. The latter is the Auger process. The energy of the ejected Auger electron is unique to the atom from which it was ejected. This energy is then detected and used to identify the atom type. The Auger process can be used to identify elements heavier than lithium (Li). An Auger energy spectrum consists of a plot of the number of electron counts, $n(E)$, of a given energy or dN/dE versus incident electron energy. The dN/dE spectrum is most widely used for thin-film coatings because the peaks can readily be separated from broad background features. Probe depths for which Auger electrons can be detected range from 5 to 100 Å.

Figure 5.6 shows the Auger depth profile of a silicon-nitride (Si_3N_4) coating.[21] Unless a milling rate is calibrated, most depth profiles plot the Auger signal [$n(E)$ or dN/dE] against ion-milling time. Sputtering time must be converted to a physical depth by appropriate calibrations (sputtering rate or yield of each coating material). Auger spectroscopy is used here to determine composition for a range of deposition conditions. Relative Si, N, and O contents are obtained from the peak intensities in the Auger-electron-energy spectrum. The average Si/N ratio is 0.75 ± 0.02

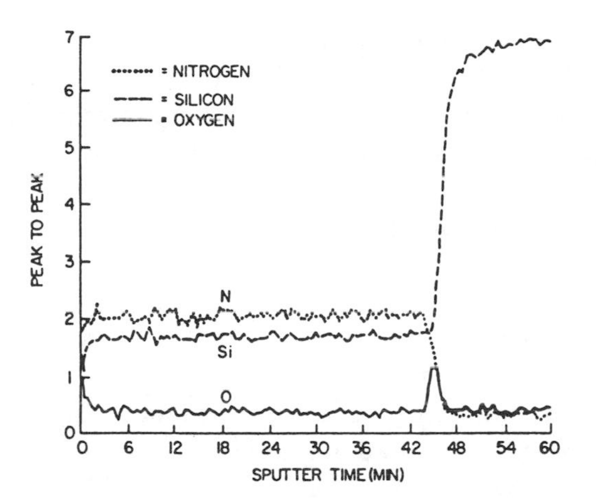

Figure 5.6 Auger depth profile of CVD silicon-nitride coating.[21]

through the entire coating thickness. A 1.5-nm-thick SiO_2 layer is found at the coating/substrate interface. In this case, Si_3N_4 and SiO_2 standards were used to calibrate both the energy and intensity spectra. The energy spectrum also demonstrates the utility of this technique for determining the composition or impurities at coating/substrate or coating/coating interfaces.

As will be discussed in Section 5.3, Multilayer Optical Coatings, because of the ion-milling depth-profiling capabilities, the Auger technique is very useful for profiling the composition of multilayer optical coatings (if charging can be eliminated).

Secondary ion mass spectrometry (SIMS) This is one of the most sensitive surface-analysis techniques and can be used to measure compositions of H to U in coatings. Secondary ions emitted by sputtering the surface are collimated into an energy filter and then collected in a mass spectrometer. SIMS can be used to obtain both surface composition and compositional depth profiles in much the same manner as AES. Surface composition profiles are obtained by rastering the ion beam across the surface of the coating. Depth profiling is achieved by sputtering into the coating while keeping the spatial position of the ion beam fixed.[15] Because the coating is sputtered, the coating is usually damaged using SIMS. The samples are placed in high vacuum during analysis.

A SIMS spectrum consists of the number of counts or secondary ions collected vs energy. Peaks in the energy spectrum correspond to the increased sputtering yield of a particular element and are unique to that specific ion mass. As with all other surface techniques, the SIMS technique must be calibrated against a known element for a given sputtering gas (usually Ar or Kr). This is a widely used technique for analyzing H concentrations and variation of H concentration in a coating.

This is a surface-sensitive technique, with an effective probing depth from 2 nm to 100 μm, depending upon sputter rate and data-collection time. Compositional depth profiling is accomplished by ion milling or sputtering the coating surface while collecting photoelectrons, as with the other surface-analytical techniques. Compositional depth profiles are also made by using two different incident X-ray energies, which excite photoelectrons with different escape energies and depths.[22] SIMS has been shown to be particularly useful in profiling the H content and distribution in semiconductor coatings.[23]

Compositional depth profiles are obtained by sputtering through the coating. Profile measurement sensitivity can be reduced by intermixing between coating and substrate or the various layers of a multilayer coating due to the sputtering and "knocking in" of ions. Atoms in the coating can be displaced and diffused into the layers below. Interpretation of the SIMS depth profile must be performed with care because this mixing results in an apparent interface broadening. The broadening is minimized by judicial choice of ion energies and sputtering conditions.

X-ray photoelectron spectroscopy (XPS or ESCA) This widely available technique involves the release of an inner-shell electron, or photoelectron, when the atom absorbs X-ray radiation. The energy of the photoelectron depends on the exact shell structure of the atom, which is characteristic of each element in the coating.[15] Most X-ray photoelectron spectrometers directly relate the kinetic energy of the electron to the binding energy of the atoms and thus to the composition of the coating. The surface of the coating is irradiated by X-rays of known energy. The important features of a photoelectron spectrum are sharp intensity peaks and extended tails of photoelectron counts. The energy-normalized signal $n(E)/E$ is plotted against binding energy (in eV). The peaks correspond to the energies of the photoelectrons which have escaped the coating with no energy loss. The asymmetry of the peaks to higher energies is due to electrons which have suffered inelastic scattering events and have an apparently higher binding energy. The peaks in the spectrum can be identified with a specific element or elements and composition. The binding energy increases with the square of the atomic number. The intensity of the peaks and integrated area under the peaks can be related to relative composition. Calibration is accomplished by using known standards and comparison with data in the literature. This technique is useful for determining the composition of elements Li through U.

If ion milling is not used, XPS is a nondestructive compositional analysis technique. Also, surface charging of insulator coatings is not a problem. Spatial resolution is poorer than that with AES or SIMS because focusing the X-ray beam is difficult.[22] Spot sizes are currently limited to about 150 μm or larger. Hydrogen cannot be detected. Analysis of XPS spectra is usually more straightforward than AES analysis because only one electron level is involved. In many cases, XPS and AES spectra can be obtained simultaneously for complementary information.

Rutherford backscattering spectroscopy (RBS) RBS is one of the most widely used compositional analytical techniques. In this process, MeV helium (^{4}He) ions are elastically scattered by the nuclei in the coating and the energy spectrum of backscattered ions is collected and stored by a multichannel analyzer. The scattering is sensitive only to the nuclear-scattering cross section and is independent of chemical bonding and electronic configuration. Because the scattering events are elastic, the target nuclei in the coating are not changed. An RBS spectrum consists of particle counts versus particle energy (stored in each channel). These data can be directly related to coating composition, atomic concentrations, and depth profiles once the system parameters (system geometry, beam current, stopping power, scattering cross section) have been defined. An excellent treatment of nuclear scattering processes can be found in Feldman and Mayer.[15]

RBS is an excellent technique to obtain simultaneous quantitative compositional depth profiles and elemental composition of elements ranging from Li to U. The energy of the backscattered ions decreases with penetration depth. The energy of the scattered particle can be related to a penetration depth. Figure 5.7 shows the RBS depth profile of a Si_3N_4 coating.[24] The profiles of Si and N taken in the 0.8-μm-thick coating were obtained using 3.4 MeV ^4He ions. The profile shows that the distribution of Si and N is relatively uniform through the entire coating thickness and that the coating is Si rich with a N/Si ratio of 1.16, which made it more optically absorbing than stoichiometric Si_3N_4.This coating also had incorporated H, which was not detected by RBS, but was detected by nuclear-reaction analysis. This demonstrates the necessity of using more than one analytical technique to determine coating composition.

The main advantages in using RBS are the exact quantitative information obtained and the rapid, nondestructive compositional depth profiling possible. This technique is insensitive to monolayers and light elements (H and He). High vacuums are also required.

Nuclear reaction analysis (NRA) The experimental arrangement for NRA and nuclear activation analysis (NAA) is much the same as for RBS. NRA, however, uses specific incident particles, incident energies, and targets. Because a nuclear reaction occurs with the inelastic collisions between particles and the coating

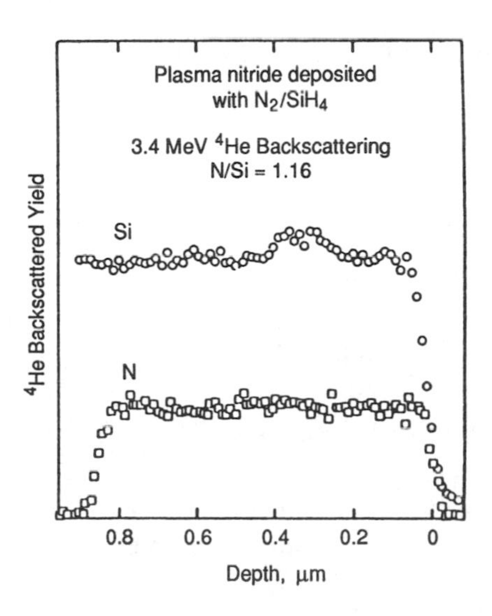

Figure 5.7 Depth profile of a silicon-nitride coating taken with RBS.[24]

nuclei, the identity of the coating material and the incident particles changes. Protons, alpha particles (^4He), and deuterons are commonly used nuclear particles. In an oxide optical coating, a proton beam will react with an ^{18}O atom to form ^{15}N and an alpha particle, $^{18}O(p,\alpha)^{15}N$.[23]

An NRA spectrum consists of the number of counts per channel of the multichannel analyzer versus energy of the emitted particle. The peaks in the spectrum correspond to the nuclear-reaction peak, which is characteristic of each type of element in the coating, and a backscattered incident particle peak. Elemental identification and compositional analysis are based on the energy position of the nuclear-reaction peak or peaks. Compositional depth profiling is accomplished by increasing the energy of the incident particle beam so that it penetrates deeper into the coating. This technique must be performed in a vacuum and is absolutely quantitative and sensitive even to H and He. Depth-profile sensitivity depends on the reaction used and has about the same monolayer sensitivity as RBS.

Some of the other surface-analytical techniques which are not as widely used and which provide much the same data as those just presented are

- energy-dispersive spectroscopy (EDS)
- electron energy-loss spectroscopy (EELS)
- electron microprobe analysis (EMA)
- extended X-ray absorption fine structure (EXAFS)
- X-ray fluorescence (XRF)
- X-ray energy spectroscopy.

Microstructural analysis TEM, SEM, electron diffraction, and XRD are the most widely used and well-established techniques for microstructural analysis of thin-film optical coatings. Microstructural analysis by TEM includes electron diffraction. STEM is a hybrid technique combining TEM and SEM analysis. In many cases, thin film coatings are ideal for these types of analyses because they are already sufficiently thin to use directly and do not require the elaborate thinning methods used to thin and replicate bulk materials. Cross-sectional TEM (XTEM) is used to obtain direct observation of microstructure and microcrystalline atomic order in single-layer and multilayer structures. Sheng[25] presents an excellent treatment of the sample preparation and replication methods for thin-film coatings. Sample preparation is destructive.

Total integrated scattering (TIS) and back reflection distribution function (BRDF) analysis are two optical-analytical techniques which assess the surface roughness and the scattering distribution of a coating using laser radiation. Because these techniques give no direct measure of microstructure, they are not addressed here.

Transmission electron microscopy (TEM) TEM is used to identify the crystalline and microcrystalline structure, grain structure, phase composition, grain boundary structure, defect structure, and interface structure of thin films. XTEM is more applicable to multilayer optical coatings, and, by using high-resolution imaging, analysts can identify lattice planes. XTEM will be discussed in more detail in the section on multilayer optical coatings. Imaging up to 500 000× can be achieved, with lattice resolution of 0.14 nm. Both dark-field and bright-field TEM images can be taken. TEM and XTEM measurements require that the sample be placed in a high vacuum. The coatings must be thin enough to transmit the electrons, but not so thin as to wash out the electron-diffraction pattern. Typical coating thicknesses for TEM range from 300 to 2000 Å. Coating thicknesses for XTEM range from monolayers to microns. In many cases, freestanding coatings are obtained by dissolving the substrate or a thin layer under the coating. In all cases, the coating is bombarded by a 50–200-keV electron beam. The electron diffraction pattern provides information about lattice spacing, lattice disorder, crystal-lattice orientation, degree of crystallinity, grain size, and crystalline phases present in the coating. TEM analysis also shows voids, defects, and grain boundary regions.

The optical properties of polycrystalline coatings, such as TiO_2, and many other multiphase systems, are determined by phase composition. The refractive indices of the anatase and rutile phases of TiO_2 are 2.3 and 2.5, respectively, and an intermediate value of n is obtained by mixing the two phases. Figure 5.8 shows a 150 000× TEM picture and the electron-diffraction pattern for a sputtered TiO_2 coating. The coating has a fine grain structure, and the electron-diffraction pattern shows that all crystallites have the rutile phase. The electron-diffraction pattern is composed of narrow rings, indicative of large grains (~1000 Å diameter), with the rutile spot pattern.

Figure 5.9 shows the dependence of n on grain size obtained from TEM analysis for pure rutile coatings.[4] The variation in n is explained by application of effective medium theory and a simple microstructural model. Coatings with the largest grains

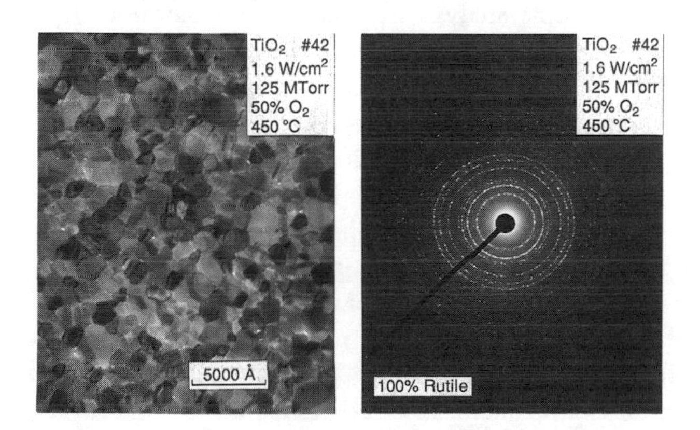

Figure 5.8 150 000× TEM picture and electron diffraction pattern from a sputtered TiO_2 coating with 100% rutile phase.

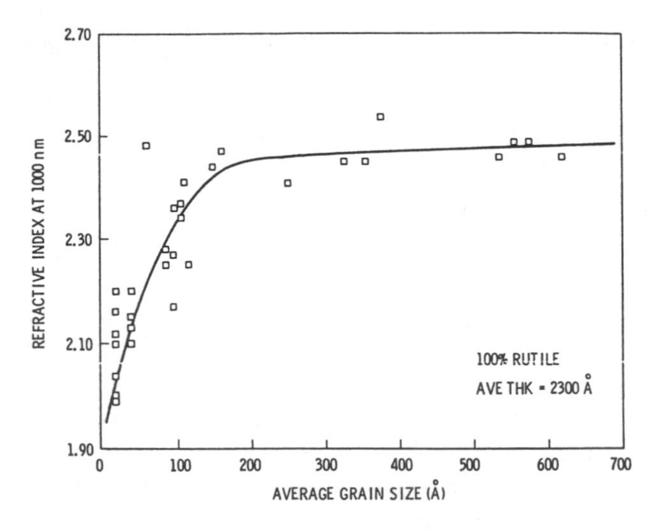

Figure 5.9 Dependence of refractive index of sputtered rutile TiO_2 coatings with grain size.

have optical properties near that of bulk rutile, and those with extremely small grains (close to amorphous) have a low *n*. The diffraction pattern of an amorphous coating would display broad fuzzy rings and no preferred crystalline-diffraction pattern. Amorphous coatings can exhibit refractive indices as low as 2.1.[4]

Scanning electron microscopy (SEM) SEM is an excellent technique to determine the surface topology of thin-film optical coatings. Determining surface morphology is important because optical scattering from the surface and the emissivity of the coating increase with surface roughness. Analysis of the surface structure can yield important information about the properties of the coating. Unlike TEM or XTEM, SEM requires no complicated sample preparation techniques and the coating is usually not damaged. Sample analysis is quick and turn-around times are short compared with TEM analysis. In this technique, a small electron beam is rastered across the surface of the sample. The magnification determines the size of the beam and ranges from 10 to 100 000. Surface features ranging from 10 to 100 nm can be resolved. Using the X-ray fluorescence capabilities, analysts can determine the chemistry of the surface featured for elements from B to U. In many cases, the coating scientist works with the microscope operator during sample analysis.

To prevent charging and loss of resolution due to bombardment of the electron beam, the surface of the coating must be conductive, or made conductive by application of a thin conductive coating (graphite or thin metal layers are commonly used). Figure 5.10 shows SEM pictures of the change of the morphology of an indium-tin oxide (ITO) optical coating with substrate temperature. The surface roughness of the coating changes from glassy to a rough-fibrous texture with an increase in substrate temperature from 50 to 450 °C. The relationship between

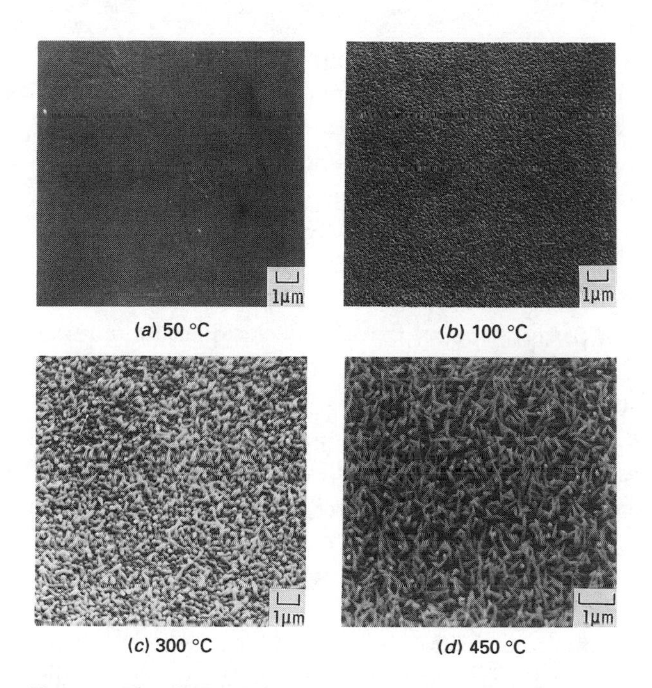

Figure 5.10 **SEM pictures at 7000× showing the dependence of surface morphology of a sputtered indium tin oxide (ITO) coating on substrate temperature.**

deposition conditions and surface morphology is important in this case in reducing optical scattering.

Scanning transmission microscopy (STEM) This technique combines features of TEM, XTEM, and SEM. The STEM can be used as a high-resolution SEM, capable of detecting about 10^{-20} g of an element, which has direct applicability to studying coating interfaces, grain boundaries, and impurities in coating defects.[26] Other applications include differential phase contrast imaging and scanning a coating to isolate small regions which can then be electron-probed. The scanning feature of STEM somewhat degrades the quality of images compared with a high-quality TEM. This, however, is more than compensated for by the scanning capability. Some modern instruments combine both TEM and STEM capabilities. STEM typically provides for spatial resolution between 0.2 and 10 nm. Quantitative measurements are difficult and still being developed, which relegates the STEM primarily to basic research. Batson[26] has written an excellent review of STEM analytical techniques.

A STEM can perform elemental X-ray analysis by the addition of an energy-dispersive X-ray spectrometer (EDS). Small compositional variations in a coating can be measured using EDS. Compositional variations at 2.5-nm distances have been measured.[27] As with TEM, STEM is insensitive to elements lighter than B.

X-ray diffraction (XRD) XRD is the most commonly used and well-developed technique to determine crystalline order and phase composition in thin-film optical coatings. Extensive reviews and books are available for reference. XRD is used to determine crystalline-phase composition, lattice constants, strains and mechanical stress, and semi-quantitative compositional analysis. XRD intensity patterns and lattice constants for many materials are well-cataloged and readily available.[28] Calibration standards are also available from the U.S. Department of Commerce, Bureau of Standards. Because X-rays must penetrate 10–100 μm into a material to obtain a good diffraction pattern, special fixturing techniques are required for analysis of thin films, which are typically less than 1-μm thick. Good diffraction patterns have been obtained for coatings as thin as 200 Å using special techniques. Often, thick coatings are fabricated specifically for XRD analysis. Counting times needed to obtain a diffraction pattern for thin-film coatings can be as long as 24 h. This technique is not effective for structural analysis of elements with $Z < 3$ (Li).

A Seemann–Bohlin diffractometer or a Read camera is used to obtain XRD patterns. Grazing-incidence XRD (glancing-angle XRD) is used to limit penetration of the X-ray beam and to enhance diffraction peak resolution by reducing the substrate diffraction pattern. A typical XRD pattern consists of intensity peaks due to Bragg reflections of X-rays plotted against diffraction angle (2θ and ϕ). Salient features are peak intensity, diffraction angle, and peak width. The lattice constants are determined from the diffraction angles at which the peaks occur. Degree of crystallinity is determined from the intensity of the peaks and the presence of peaks corresponding to specific phases. Amorphous coatings, for example, have very weak and broad diffraction patterns. Average grain size is determined from the line width at half maximum. The integrated area under the diffraction peak is directly related to the relative phase composition.

If a Read camera is used, the diffraction pattern is an ordered array of intense spots and rings. The spots and rings result from Bragg reflections off the coating. The angular positions of the spots and rings correspond to a specific phase composition. The diffraction patterns are compared with American Society for Testing and Materials powder pattern compilations to identify the crystalline phase composition.

5.3 Multilayer Optical Coatings

The optical performance of a multilayer optical coating depends on the optical properties of each individual layer, which in turn depend primarily on composition. The coating designer and engineer must be aware of how the composition and microstructure change *n* and *k*, the wavelength dispersion of each thin film-layer material used in the multilayer coating, and the resulting optical performance. The optical performance of the multilayer can be completely modeled once *n*, *k*, and layer thicknesses are known, or the coating design can be determined given *n*, *k*, and target reflectance and transmittance. Small deviations in composition or microstructure can alter coating performance from design specifications. For example, for the same number of high/low refractive-index pairs, the peak reflectance and full width at half maximum of a quarterwave high-reflector coating increase with increased refractive-index ratio (n_H/n_L) and decrease with decreasing n_H/n_L. The peak reflectance decreases with increased k. If a small compositional change occurs which decreases n_H (or increases n_L), the peak reflectance and the bandwidth of the coating will decrease, and the performance may not meet design specifications.

Deviation in the performance of multilayer coatings from the design performance is the first clue that layer composition may be incorrect, or may have changed, and that the deposition process requires recalibration. Compositional changes which make a coating metal-rich, or substoichiometric, usually increase both n and k. This behavior is typical of SiO_2 and Si_3N_4 and of most oxide and nitride coatings. In these cases, the optical band edge also shifts to longer wavelengths (lower energy), which may cause a problem for coatings designed to operate at short wavelengths. Both n and k decrease with the incorporation of H or F,[29] and physisorption of water due to porosity and voids causes significant changes in n and k. This behavior is typical of many coatings deposited by low-energy processes, such as evaporation.

This section addresses the applicability of the analytical techniques presented for single-layer optical coatings to the compositional and microstructural analysis of multilayer optical coatings. Many of the single-layer analytical techniques do not supply satisfactory information for analysis of multilayers, while some techniques are ideally suited to multilayer analysis.

Compositional Analysis

The same analytical techniques presented for single-layer optical coatings are addressed in this section. The coating engineer must, however, be knowledgeable of how layer composition and changes in the layer compositions affect optical performance of the completed multilayer coating and what analytical techniques work best for each coating material.

Infrared spectrophotometric analysis This technique is better suited for quantitative compositional analysis of single-layer than multilayer coatings for the following reasons: (1) the presence of two or more coating-material types severely complicates the interpretation of optical transmission and absorption spectra; (2) optical interference effects can either enhance or wash out molecular vibrational features and complicate their measurement; (3) interfacial layers may add unwanted vibrational features (chemisorbed water, hydrocarbons, and impurities) and further complicate analysis; and (4) the composition of each layer of the same desired composition may be slightly different if the deposition process parameters drifted during the coating run. However, a qualitative analysis to determine the types of molecular bond configurations in the coating can be obtained. Other analytic techniques which can depth-profile coating composition, such as AES, RBS SIMS, and NRA, should be chosen if available.

The center wavelengths of IR-vibrational-absorption bands characteristic of the local bond configurations can still be used to identify the various types of molecular bonds in the coating. Care must be taken in quantifying the relative amounts of each bond type. If the exact thicknesses, the refractive indices, and extinction coefficients of all the layers are known, the intensity spectrum of the absorption bands can be calculated by comparison with the spectrum expected with no absorption.

This technique is excellent for detecting the presence of impurities and physisorbed water in the multilayer coating, although it will not indicate which layer is affected or if interfaces between layers are affected. For example, a strong absorption near a wavelength of 2.94-μm which results from water physisorbed in the coating due to porosity, interface defects, grain boundaries, or particulates is easily detectable using optical techniques. This absorption severely degrades optical performance in the NIR and is enhanced in multilayer coatings due to the increased number of sites to take on water. In some cases, impurities would not be detected by spectrophotometric analysis. Oxygen contamination in the Si_3N_4 high-index layers of a Si_3N_4/SiO_2 multilayer would be difficult to detect due to the SiO_2 low-index layers. Oxygen or water contamination of nitride layers in a multilayer stack usually shows up as a reduced peak reflectance and reduced bandwidth at half maximum of a high reflector, or as an increased reflectance of an antireflection coating. Alternate compositional analytical techniques must be used in cases of ambiguity and insensitivity of IR spectrophotometric analysis.

Laser Raman spectroscopy Laser Raman spectroscopic techniques are well-suited for determining the relative phase composition of optical multilayer coatings. High-quality Raman spectra can be obtained to determine the relative phase composition of the individual layers usually without special enhancement techniques. In many cases, the spectra of the individual layers are enhanced by the multilayer design. Signal enhancement is achieved by multiple reflections and passes of the probe radiation in the coating. Because the spectrum of the coating is intensified due to interference effects, the substrate spectrum is reduced in relative intensity and becomes less of a problem. The Raman peak intensities of the spectra are modulated by the transmission spectrum of the multilayer stack. The degree of signal enhancement depends on the probe laser wavelength and the multilayer coating's transmission of the probe wavelength.[30] If the probe wavelength is in a high-reflectance band, little or no signal enhancement is observed. Probe wavelengths centered on sidebands near the primary reflection peaks provide signal enhancements up to an order of magnitude compared to single-layer coatings. Efficient use of this technique to enhance the Raman signal requires a detailed study of the electric-field distribution in the multilayer and a laser with multiple-probe wavelengths.[31]

Surface Analytical Techniques

The depth profile capability of the surface analytical techniques discussed in the section "Surface Analytical Techniques" makes them well-suited for determining compositional thickness profiles of multilayer optical coatings. Because all the surface analysis techniques used ion milling or sputtering to obtain a compositional depth profile, their application to multilayer optical coatings are presented together. The nuclear analytical techniques, RBS and NRA, also are presented together.

Compositional depth profiling of optical coatings with more than two or three layers appears not to be a common practice. These techniques are often used to determine the degree of diffusion between layers, impurities in layers and interfaces, and chemical attack such as oxidation at interfaces, grain boundaries, and the coating surface. Historically, because they can detect low levels of impurities, they are currently finding more applications in the multilayer semiconductor analysis than in compositional analysis of optical coatings. AES and other techniques tend to lose resolution and sensitivity with increased profile depth, although excellent work has been done with GaAs and AlGaAs multilayer quantum wells.[20] Depth resolution is often lost due to preferential sputtering of the different layers in the multilayer stack and to "knock-in" effects which mix layer materials and smear out interfaces. Most compositional profiles do not go beyond two or three layers, which is usually typical of the coating.

Both RBS and NRA techniques are directly applicable in determining the composition or composition gradients of multilayer coatings. RBS is an excellent technique to study composition, layer interdiffusion, impurities at the substrate-coating interface, and surface chemical attack (such as oxidation and sulfidation) of multilayer coatings. Concentration levels between 10^{-4} and 10^{-2} are routinely detected. As with the other surface techniques, however, RBS has been used extensively with thin-film semiconductors, and most recently with thin-film superconductors; little is reported for multilayer optical coatings. NRA is used to detect concentration, concentration gradients, and diffusion or contamination of light elements from hydrogen to silicon with about the same sensitivity as RBS. Hydrocarbons between layers and at the substrate interface which result from vacuum system impurities can be profiled.

Microstructural Analysis of Multilayer Optical Coatings

All techniques used to determine coating microstructure discussed in the section "Microstructural Analysis" are also extensively used to determine the microstructure of multilayer optical coatings. Sectioning and thinning techniques to obtain a cross section of the coatings are well developed and provide excellent replication of the multilayer microstructure.[32] These techniques have reached a degree of sophistication which enables the study of microstructural features as small as crystalline-lattice matches, or mismatches at boundaries of epitaxial films. The microstructural zone classifications[33] can be assigned to the coating layers and related to deposition conditions. This section highlights state-of-the-art microstructural analysis by TEM, STEM, SEM, and XRD.

These techniques are often used in conjunction with optical measurements to determine how the coating microstructure has improved or degraded optical performance. Types of microstructure which can degrade multilayer-coating performance are porosity, columnar structure, polycrystallinity, wetting gaps at interfaces, less dense grain-boundary structure, and particulates in the layers and interfaces.

TEM and STEM Because TEM, STEM, and electron diffraction analyses are often performed by the same instrument, and because these techniques provide much the same type of information, their application to multilayer optical coatings is discussed together. Additional information is required and can be obtained for multilayer coatings compared to single layers. In addition to crystalline phase composition, grain size, density, defect density, and columnar structure, the microstructural changes at interfaces, interface density and microstructure, lattice plane continuity, layer interdiffusion, and uniformity of layer thicknesses can be determined. Figure 5.11 shows a 150 000× TEM cross section of an $Al/(SiO_2/Si_3N_4)$ laser reflector made by magnetron sputtering. The Si_3N_4 layer (*top*) displays columnar features approximately 800 Å in diameter, although electron diffraction showed it to be amorphous. Voids with 150-Å diameter are present in the SiO_2 layer (*middle*). Again, electron diffraction showed this layer to be amorphous.

Microstructural analysis such as that shown in Figure 5.11 can often be used to guide deposition process changes which improve coating microstructure and optical performance. In this case, the columnar structure of the Si_3N_4 layer caused a refractive index slightly lower than expected. The porosity of the SiO_2 layer caused slight reflectance-peak shifts due to moisture pickup when exposed to air. Both indicate that process changes are needed to densify the coatings and reduce particulates.

SEM SEM is used extensively to evaluate multilayer-coating surface morphology. Although techniques have been developed to smooth the surface roughness of layers of a multilayer stack,[34] the surface of the completed stack is usually rougher than the roughest layer of the stack. Surface roughness, which causes optical

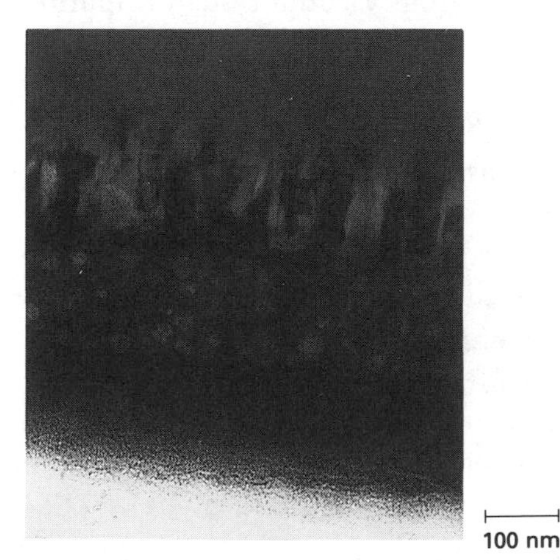

Figure 5.11 TEM picture at 150 000× of a multilayer silicon-nitride/silicon-oxide laser reflector coating.

scattering, results from porosity, polycrystallinity, and particulates in the coating. Coating layer roughness features either replicate or enlarge with increased layer numbers and thickness. As with single-layer coatings, the same type of surface chemical analysis is performed by energy-dispersive spectroscopic analysis, to determine the composition of surface features as small as 10 nm. Note that, as with single-layer coatings, the surface of the multilayer coating must be made conductive to prevent charging of the electron beam.

XRD Analysis of multilayer coatings by XRD is essentially the same as that of single-layer coatings discussed in the subsection entitled "X-ray diffraction (XRD)." Care must be taken to attribute the correct phase composition to each layer, again guided by single-layer results. The phase composition of all layer materials of a coating can be obtained from one XRD spectrum. In most cases, this should be straightforward because the possible crystalline phases of the layer materials with different compositions will likely be different. For example, a TiO_2 /SiO_2 multilayer will display the TiO_2 anatase or rutile peaks, and the SiO_2 will display broad peaks of the hexagonal structure. Both may be amorphous or polycrystalline, or one layer may be amorphous and one polycrystalline. Grain sizes can be obtained from broadening of the diffraction peaks. Unlike laser Raman spectroscopy, with XRD the interference effects are negligible.

XRD can be used in conjunction with Raman spectroscopy, RBS, NRA, AES, or the other surface-analytical techniques to provide positive identification of compounds formed by diffusion and mixing of layers. This is accomplished by identifying the crystalline phase or phases present and the corresponding chemical alloy or mixture which possesses that specific phase from phase/composition diagrams.[15]

5.4 Stability of Multilayer Optical Coatings

Irreversible and undesirable changes in the optical performance and physical integrity of a multilayer optical coating can be caused by atmospheric moisture, chemical attack, interdiffusion between layers, poor adhesion, particulates, and high temperatures. The stability of the multilayer coating is directly related to the compositional and microstructural stability of its layers. Many times the instability of one layer material can degrade the performance of the entire coating. The compositional and microstructural analytical techniques presented in this chapter can be used to determine the mechanisms for instability. Techniques such as TEM, STEM, XRD, and Raman spectroscopy are especially useful in the evaluation of coating instability. Raman spectroscopy, in particular, is used in situ to evaluate changes in phase composition. Choice of technique, however, will depend on availability and coating composition. For example, Figure 5.12 shows how physisorbed water due to a porous microstructure of the Al_2O_3 layers of a Si_3N_4/Al_2O_3 high reflector severely degrades the peak reflectance of the coating at the design wavelength near 2.8 μm. The reflectance is severely reduced by an absorption band near 2.94 μm due to the physisorbed water. Water in the coating was detected by IR spectroscopic

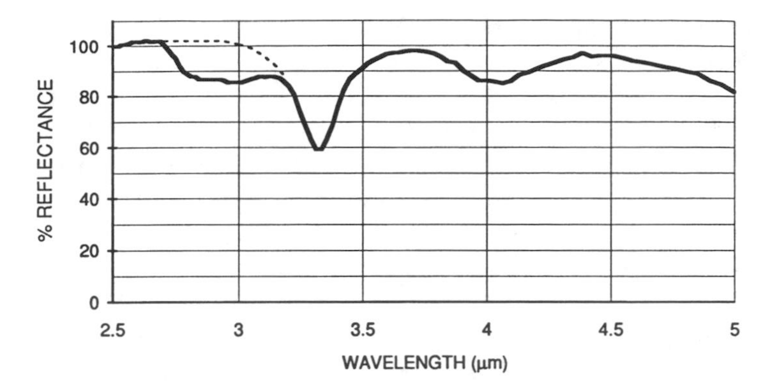

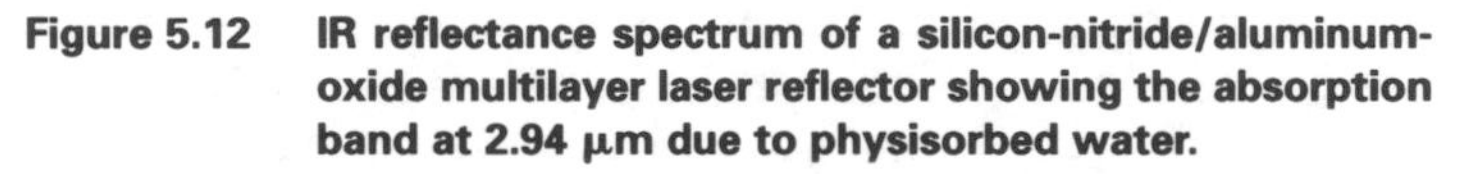
Figure 5.12 **IR reflectance spectrum of a silicon-nitride/aluminum-oxide multilayer laser reflector showing the absorption band at 2.94 μm due to physisorbed water.**

analysis, and TEM analysis showed a porous microstructure. In addition to undesired absorptions, the change in *n* due to water incorporated into a coating causes shifts in the reflectance peak, transmission notch, or filter edge, often by as much as 15 nm. TEM and STEM analysis of this coating showed a porous microstructure of the Al_2O_3 layers and a columnar microstructure of the Si_3N_4 layers similar to that shown in Figure 5.11. Improvements in the optical performance and stability due to modifications in the deposition process were monitored using these techniques.

A multilayer coating with fully dense and stoichiometric layers is usually very stable (unless the bulk materials are intrinsically unstable) and much less susceptible to attack. The reflectance of the same coating design as shown in Figure 5.12, when made with denser Al_2O_3 layers, displayed little or no water-related absorption.

Many times a particular phase or composition is more unstable than another. Microstructural and compositional analyses are required to determine if the more stable phase or composition has been achieved. The hexagonal phase of BN is readily attacked by atmospheric moisture, while the cubic phase is extremely hard and moisture-resistant. Substoichiometric BN is also susceptible to moisture attack. Phase composition of a coating can also be stabilized by addition of small amounts of another material. Small amounts of Y_2O_3 are often used to stabilize the high-temperature cubic phase of ZrO_2. Correct phase compositions can be verified by XRD, EELS, Raman spectroscopy, and other techniques.

The stability and durability of a multilayer coating is often improved by application of a protective-overcoat layer incorporated into the coating design or by replacing unstable layer materials with fully dense stable-layer materials. Protective coatings have received extensive treatment in the literature. Undercoat layers are used to improve the transfer of heat energy from the coating to the substrate or to diffuse heat absorbed by the substrate. Here again, TEM and STEM can be used to verify that the overcoat is dense and free of defects (such as pin holes) and that the undercoat has the phase composition which gives good thermal conductivity.

5.5 Future Compositional and Microstructural Analytical Techniques

Several promising new techniques for compositional and microstructural analysis are currently being developed, but have not reached the state of maturity and the reliability required for day-to-day use and commercialization. Four of the most interesting techniques are photoacoustic spectroscopy (PAS), scanning acoustic microscopy (SAM), picosecond acoustic analysis, and medium-energy backscattering spectrometry. These techniques will not supplant those already presented, but will provide additional detail and more localized compositional and microstructural information. All four techniques are nondestructuve.

PAS, SAM, and picosecond acoustic analysis are related techniques which detect the acoustic signal from absorption-induced heating in a coating. In PAS, the coating is "pumped" by an intensity-modulated light source. Part or all of the optical excitation is converted to heat in the coating due to absorption. The acoustic signal is detected by a piezoelectric transducer. The acoustic signal can be directly related to density, bonding, defects, surface and interface composition.[35] Depth profiling has been achieved.

SAM is a related technique which has demonstrated a variety of unique capabilities in imaging and characterization of compositional, surface, subsurface, and interfacial details.[36] Spot sizes as small as 10 μm have been used to profile coating homogeneity. Voids and defects have also been imaged in coatings with improved resolution compared to X-ray imaging.

Picosecond acoustic analysis is used to determine the thickness, homogeneity, and bonding in coatings with thickesses ranging from 0.05 to several microns.[37] A nanosecond laser pulse, focused onto the coating, produces a thermal stress, and a longitudinal acoustic wave. The spatial dependence of the reflected acoustic pulse is related to bonding of the coating at the substrate and to homogeneity of the coating. This technique has been used with semiconductor optical coatings and oxides.

Medium-energy backscattering spectrometry combines the quantitative analysis of RBS with inceased surface sensitivity, depth resolution, and reduced coating damage.[38] This technique employs He^+ and Li^+ ions with energy ranging from a few tenths of keV to 500 keV in a time-of-flight spectrometer. A high-energy particle accelerator is not needed. Quantitative information and compositional depth profiles with 10% accuracy can be obtained for elements ranging from Be to U. Typical concentrations which can be detected are about $10^{15}/cm^3$ for Be to $10^{13}/cm^3$ for Au.

References

1 P. M. Martin and W. T. Pawlewicz. *Solar Energy Materials.* **2**, 143, 1979/1980.

2 Y. Katagiri and H. Ukita. *Appl. Optics.* **29** (34), 5074, 1990.

3 H. A. Macleod and D. Richmond. Thin Solid Films. **37**, 163, 1970.

4 W. T. Pawlewicz, P. M. Martin, D. D. Hays, and I. B. Mann. *Proceedings.* Vol. 325, SPIE, 1982, p. 105.

5 *Handbook of Spectroscopy.* Vol. 2. (J. W. Robinson, Ed.) CRC Press, Cleveland, 1974.

6 G. M. Barrow. *Introduction of Molecular Spectroscopy.* McGraw-Hill, New York, 1962.

7 R. P. Bauman. *Absorption Spectroscopy.* Wiley, New York, 1962.

8 M. H. Brodsky, M. Cardona, and J. J. Cuomo. *Phys. Rev. B.* **16**, 3556, 1977.

9 B. Dischler, A. Bubenzer, P. Koidl, and G. Brandt. *Proceedings.* Vol. 400, SPIE, 1983, p. 122.

10 W. A. Lanford and M. J. Rand. *J. Appl. Phys.* **49** (4), 2473, 1978.

11 A. H. Lettington, C. J. H. Wort, and B. C. Monachan. *Proceedings.* Vol. 1112, SPIE, 1989, p. 156.

12 M. LeContellec, J. Richard, A. Guivarc'h, E. Ligeon, and J. Fontenille. *Thin Solid Films.* **58**, 407, 1979.

13 G. J. Exarhos. *Thin Film Optical Coatings III.* (K. H. Guenther, Ed.) Springer-Verlag, Berlin, 1993.

14 W. T. Pawlewicz, G. J. Exarhos, and W. E. Conway. *Appl. Optics.* **22** (12), 1837, 1983.

15 L. C. Feldman and J. W. Mayer. *Fundamentals of Surface and Thin Film Analysis.* North-Holland, New York, 1986.

16 N. S. McIntyre. *Practical Surface Analysis.* (D. Briggs, Ed.) John Wiley and Sons, Chichester, U.K., 1983, p. 397.

17 *Practical Surface Analysis by Auger and X-ray Photo Electron Spectroscopy.* (D. Briggs and M. P. Seah, Eds.) John Wiley and Sons, Chichester, U.K., 1983.

18 *Ion Beam Handbook for Materials Analysis.* (J. W. Mayer and E. Rimini, Eds.) Academic Press, New York, 1977.

19 *Annual Book of ASTM Standards,* Vol. 03.06. American Society for Testing and Materials, Philadelphia.

20 *Thin Films: Interdiffusion and Reactions.* (J. M. Poate, K. N. Tu, and J. W. Mayer, Eds.) Wiley & Sons, New York,1978.

21 P. Pan, W. Berry, I. Assur, and S. Burton. *Proceedings.* Vol. 83-8. The Electrochemical Society, 1983, p. 88.

22 D. R. Baer and L. S. Dake. "Oxidation of Metals and Associated Mass Transport." *Proceedings.* AIME, Orlando, FL, 1986, p. 185.

23 C. W. Magee and D. E. Carlson. *Proceedings.* Vol. 78-3. The Electrochemical Society, 1978, p. 151.

24 P. S. Peercy and H. J. Stein. *Proceedings.* Vol. 83-8. The Electrochemical Society, 1983, p. 3.

25 T. T. Sheng. "Cross-Sectional Transmission Electron Microscopy." In *Analytical Techniques for Thin Films.* (K. N. Tu and R. Rosenberg, Eds.) Academic Press, Boston, 1988.

26 P. E. Baston. "Scanning Electron Microscopy." In *Analytical Techniques for Thin Films.* (K. N. Tu and R. Rosenberg, Eds.) Academic Press, Boston, 1988.

27 A. J. Garrett-Reed. *Proceedings.* (Bailey, Ed.) 41st Electron Microscopy Society of America, San Francisco Press, 1983, p. 374.

28 ASTM Bulletin No. 160, American Society for Testing and Materials, 1949.

29 P. M. Martin. *J. Electrochemical Society.* **130**, 6, 1983.

30 G. J. Exarhos and W. T. Pawlewicz. *Appl. Optics.* **23** (12), 1986, 1984.

31 R. A. Craig, G. J. Exarhos, W. T. Pawlewicz, and R. E. Williford. *Appl. Optics.* **26**, 4193, 1987.

32 J. M. McCarthy, L. E. Thomas, W. T. Pawlewicz, W. S. Frydrych and G. J. Exarhos. In *Proceedings.* Materials Research Society Meeting, San Diego, 1989.

33 B. A. Movchan and A. V. Demchishin. *Phys. Met. Metallogtr.* **28**, 83, 1969.

34 R. McNeil, L. J. Wei, G. A. Al-Jumaily, S. Shakir, and J. K. Mclver. In *Proceedings.* Topical Meeting on Optical Interference Coatings, Monterey, CA, 1984, pp. ThA–A4.

35 *Applied Optics.* **21** (1), 1982.

36 C. C. Lee, G. Matijasevic, X. Cheng, and C. S. Tsai. *Thin Solid Films.* **154**, 207, 1987.

37 C. Thomsen, H. J. Maris, and J Tauc. *Thin Solid Films.* **154**, 217, 1987.

38 M. H. Mendenhall and R. A. Weller. *Proceedings.* Vol. 1323. SPIE, 1990, p. 299.

6

Characterization and Control of Stress in Optical Films

BRADLEY J. POND

Contents

6.1 Introduction

Essentially all vacuum-deposited films are in a state of stress. In many cases the stress is high enough that it dictates the maximum thickness of a film or the combinations of materials which can be used together in making multilayer films. Film stress also limits the applications of many materials. For example, magnesium fluoride (MgF_2) has many desirable properties such as a low refractive index, good durability, and a broad transparency range. However, the high tensile stress of this material limits its applications to thin coatings such as visible antireflection coatings. In thick multilayer coatings, materials with opposing stresses are usually used to reduce the total stress of a coating. This technique was developed soon after the introduction of optical coatings since it is essential for the production of coatings of any substantial thickness. Unfortunately, there are a limited number of material combinations with opposing stresses because almost all evaporated films have tensile stresses. Commonly used material combinations which have opposing stresses include titania/silica (TiO_2/SiO_2), zirconia/silica (ZrO_2/SiO_2), zinc sulphide/cryolite (ZnS/Na_3AlF_6), and zinc selenide/thorium fluoride ($ZnSe/ThF_4$). Even when materials with opposing stresses are used, coatings still have a net stress either compressive or tensile. In thick coatings this net stress can cause distortion of the substrate or even mechanical failure of the film if the stress is high enough. Film stress is more of a problem in thick coatings because the strain energy which causes substrate distortion or mechanical failure is dependent on both film stress and film thickness.

The mechanical failure of a film due to stress is occasionally witnessed by technicians unloading substrates from a hot coating chamber. As the substrates cool and adsorb moisture from the atmosphere, the thermal and adsorption-induced stresses can combine with a film's intrinsic stress to cause a coating which looked perfect in the coating chamber to fail mechanically by delaminating. Thermal stresses are due to differences in the expansion coefficients of the film and substrate. Adsorption-induced stresses are due to the adsorption and desorption of moisture by porous coatings.[1] Intrinsic stresses include all stresses which cannot be attributed to thermal mismatch. The intrinsic stress of almost every evaporated coating is tensile and is explained in terms of a constrained relaxation during film growth from a disordered state to a more ordered state. This change results in a decrease in volume and the formation of tensile stresses.[2] In contrast to that of evaporated films, the stress of most sputtered films is compressive.[2] For example, films of MgF_2 deposited by evaporation typically have a high tensile stress[1] of 47×10^8 dyne/cm^2, whereas the same material deposited using ion-beam sputtering has a compressive stress[3] of 30×10^8 dyne/cm^2. The compressive stress of sputtered films is caused by the constant bombardment of the growing film by energetic species, which is referred to as atomic peening.[2] The total stress values for optical coating materials vary significantly depending upon a number of factors, but most materials have a compressive or tensile stress on the order of 5×10^8 to 30×10^8 dyne/cm^2.

Just as the film stress is dependent on the deposition process, it is also strongly dependent on the material being deposited and the deposition parameters. The dependence of stress on deposition parameters is often used to control the stress in coatings. In some cases the total film stress can be changed from compressive to tensile by adjusting a single parameter.[1] Unfortunately, this sensitivity of film stress to deposition parameters often makes it difficult during the production of thin film coatings for engineers to control the stress accurately from one run to the next. Alternate techniques have also been developed for modifying film stress, including the addition of impurities to a film,[4] post-deposition baking,[5] and the use of ion-assisted deposition (IAD).[6]

A variety of techniques has been developed for measuring film stress. These vary from measuring the distortion of a coated part with a simple geometrical optic setup to interpretation of X-ray diffraction (XRD) measurements. The most commonly used stress measurement technique involves measuring the amount that a thin disk is deformed by the stress of a coating.[7] In this technique, the stress is usually measured after the witness has been removed from the chamber so that any thermal stress component or any adsorption-induced effects will be included in the measured value. A cantilevered beam technique is also based on the distortion of a thin substrate, but this technique is generally used for in situ measurements.[7] This allows one to measure just the intrinsic stress component. Techniques based on the bending of a substrate can provide an accurate measure of the film stress, but this is the only property they measure. Other techniques such as XRD and Raman spectroscopy measure other properties in addition to stress. Specifically, XRD can measure the crystalline phase and stress of polycrystalline materials.[8] In addition, XRD can measure the macrostress, microstress, and stress anisotropy of thin films, whereas substrate deformation techniques only measure the macrostress of a thin film. Raman spectroscopy is limited in the number of materials which can be analyzed, but the technique is versatile and can measure a number of film properties, including stress.[9]

The remainder of this chapter presents a brief discussion of the aforementioned topics: the origins of stress in films, methods that can be used to control or modify film stress, and techniques for measuring film stress. The reader is referred to the references at the end of this chapter for more detailed information on these subjects. It should be pointed out that there is an inconsistency in the sign convention used in the literature to indicate the stress state for stress values reported. A survey of the literature shows that the majority of authors use a positive value to indicate tensile stress and a negative value to indicate compressive stress. This convention is adopted in the text of this chapter, although there is some inconsistency in the sign convention used in the figures in this chapter. Also, a variety of units are used in reporting stress values. The conversions among the commonly used units are given in Table 6.1. Stress values given in references are converted into dynes/cm^2 when discussed here.

6.2 Origins of Stress

Normally, stress in a material originates from an external force compressing or expanding the atoms in a lattice into a space of a size different from what they would occupy in a relaxed state. In thin films, intrinsic stresses are generated during

	TO					
FROM	**KBar**	**KPSI**	**MPa**	**Kg/cm^2**	**10^8 Newton/m^2**	**10^8 Dyne/cm^2**
KBar	1	14.5	100	10.2	1	10
KPSI	6.89 E–2	1	6.9	0.70	6.89 E–2	0.689
MPa	0.01	0.145	1	0.10	0.01	0.1
Kg/cm^2	9.8 E–2	1.42	9.8	1	9.8 E–2	0.98
10^8 Newton/m^2	1	14.5	100	10.2	1	10
10^8 Dyne/cm^2	0.1	1.45	10	1.02	0.1	1

Table 6.1 Stress conversion table. Multiply by the factor shown in the table to convert from one unit to another.

film growth. The spacings of atoms in a growing film are either smaller or larger than the natural spacing, resulting in compressive or tensile stress. This stress generates enough of a moment to deform the surface of a substrate. A compressive film deposited on a flat substrate will deform the surface to a convex shape, and a tensile film will deform the surface to a concave shape.

In some cases, film stress can cause the film to fail mechanically. This occurs when the strain energy stored in the film exceeds the adhesion energy. Strain energy in a film is dependent upon both the stress and the thickness of a film. An example of a film which has failed mechanically due to an excessively high compressive stress is shown in the micrograph in Figure 6.1*a*. A portion of this film has failed by buckling and delaminating, which is characteristic of a compressive stress failure. This type of failure usually moves across a substrate surface at a relatively slow speed, which can be easily observed and documented.[10] The shape of the wrinkles in this example is also characteristic of a compressive stress failure. Films with high tensile stress tend to fail in an alternate manner. These films typically fail by fracturing or crazing. This is illustrated in Figure 6.1*b*. The micrograph shows that the fractures tend to originate at defects in the coating, which is due to the stress concentration in the region of point defects.[11] Fractures usually start at these point defects and then propagate. The micrograph also shows that the fracture or craze lines always intersect at right angles and can be distinguished from scratches in the substrate in this manner. Even though the film has failed by fracturing, the film remains adhered to the substrate and is still useful for some applications. If the strain energy due to the film's stress and thickness exceeds the adhesion energy, then the film will fail by delaminating.

The total stress of a thin film is composed of a thermal stress and an intrinsic stress. The thermal stress is due to the difference in the thermal expansion

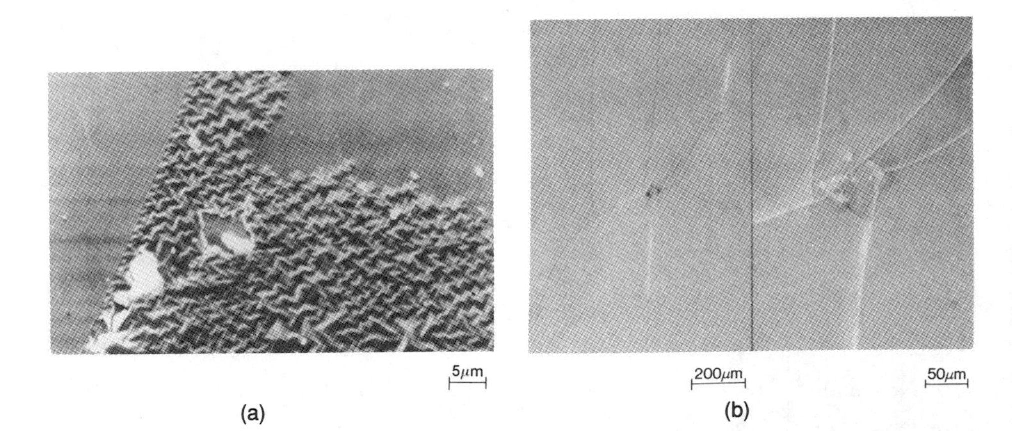

Figure 6.1 **Mechanical failure of optical thin films due to excessive stress: (*a*) compressive buckling waves in SiO_2 thin film and (*b*) tensile fracture in a TiO_2/SiO_2 multilayer coating.[10]**

coefficients of the coating and substrate materials. Most evaporated films have a thermal stress component since they are typically deposited at elevated temperatures. The thermal stress induced in a film is given in a one-dimensional thin-film approximation by

$$\sigma_{\text{thermal}} = E_f(\alpha_f - \alpha_s)(T_d - T_a)$$

where E_f is Young's modulus of the film, α_f and α_s are the thermal expansion coefficients for the film and substrate, T_d is the substrate temperature during deposition, and T_a is the ambient substrate temperature during measurement.[12] A positive value indicates the stress is tensile. For accurate calculations, measured film values should be used in this equation for the thermal expansion coefficient and Young's modulus. Experiments have shown that these parameters for evaporated films can differ by a factor of 2 to 10 from bulk values.[13] In most cases, the thermal stress is only a small part of the total stress, but in some cases it can be large enough to cause the failure of a film. This is especially true with modern applications in which the substrates can vary from low-expansion glasses that have coefficients of thermal expansion (CTEs) as low as 0.03×10^{-6}/K to plastics which have CTEs as high as 95×10^{-6}/K. Dielectric films have CTEs on the order of 1 to 30×10^{-6}/K. An example of the variation of the thermal stress on various substrates is illustrated in Figure 6.2. In this example,[14] a single-layer film of ThF_4 is deposited using electron-beam evaporation onto substrates heated to 200 °C. The film was simultaneously deposited on a low-expansion glass (Cer-Vit™) and on potassium chloride (KCl), a substrate with a high thermal expansion coefficient. Film stress was monitored in situ both during film growth and while the substrates were cooled to room temperature. In the initial stages of film growth the stress increased rapidly, and at a particular thickness the change in film stress decreased and became nearly constant. This is typical of most vacuum-deposited films. After deposition stopped and substrates had been cooled to room temperature, the effects of the thermal stress component became apparent. The thermal stress component for the film deposited on the Cer-Vit™ is tensile because the film shrinks more than the substrate upon cooling. In general, films deposited on low-expansion glasses such as Cer-Vit™, Zerodur™, or Ultra Low Expansion (ULE™) have a tensile thermal stress component because evaporated films have CTEs 10–1000 times greater than those of the low-expansion glasses.[13] For the film on the KCl substrate, the film has a lower expansion coefficient than the substrate and the thermal stress component is compressive. In this case, the thermal stress is greater than the intrinsic stress and the total film stress is compressive. This effect is also seen in substrate materials such as plastics or metals, which have high CTEs in comparison to those of dielectric films.

Intrinsic stress is the component of the total stress that cannot be attributed to thermal mismatch. Almost all evaporated coatings have tensile intrinsic stresses, and several models have been developed to explain the origins of these stresses. Two

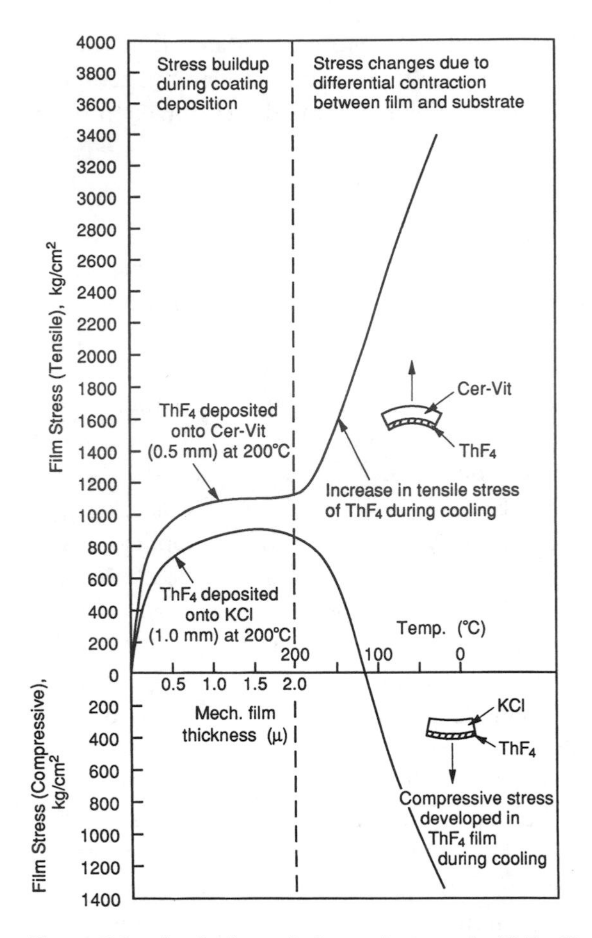

Figure 6.2 **Intrinsic and thermal stress in ThF_4 films deposited onto Cer-Vit™ and KCl substrates at 200 °C followed by cooling to ambient temperature.[14]**

models are discussed here, both of which employ a constrained relaxation or constrained shrinkage mechanism during film growth to explain the tensile stress. The first model assumes that the outer surface layer of a growing film is in a condition of high disorder, far from thermodynamic equilibrium. The regions beneath the surface layer relax toward a state of increased order and decreased volume. The decrease in volume results in the formation of tensile stresses.[2] The second model is known as the grain boundary model.[15] In this model, the crystallites in the growing film are visualized as growing together until the gap between the two crystallites is on the order of the bulk lattice constant of the film material. The interatomic forces then produce an elastic relaxation of the boundaries toward each other. This relaxation is constrained because of the crystallites' adhesion to the surface. This constrained relaxation results in the formation of tensile stresses. A columnar growth simplifies the grain boundary model, an approach commonly used in studies of intrinsic stress. In this approach the tensile stress is generated by the grain boundary energies of parallel columnar grain boundaries.

There has been only limited study of the origins of intrinsic stress in evaporated optical coating materials. The grain boundary model has been successfully used to explain the intrinsic stress in MgF_2 and ZnS.[1] In this study the grain boundary model is interpreted in terms of the forces at the columnar grain boundaries. Calculations of the tensile stress of MgF_2 agree very well with the measured values. This study also shows that impurities in MgF_2 films tend to congregate at the grain boundaries and cause considerable stress reduction. The presence of impurities at the grain boundaries has been used to explain the compressive stress of the ZnS films. It was also shown that the adsorption of moisture can reduce the tensile stress of MgF_2 films. This is due to a change of the surface energy of the columnar grain boundaries. Additional studies of moisture adsorption effects on film stress have shown that the effect is material-dependent. The adsorption of moisture increases the compressive stress of SiO_2 films, while it increases the tensile stress of TiO_2 films.[16]

The intrinsic stress of most sputtered films is compressive, and these stresses are explained in terms of an atomic peening mechanism.[2] In a sputtering process, the sputtered species are ejected from the target with an average energy of tens of electron volts, whereas evaporated species have an average energy of a few tenths of 1 eV.[17] Thus, the average energy of the species striking the growing film is much higher for sputtered films. In addition, the growing film is bombarded by energetic neutrals.[18] These are accelerated ions that are neutralized at the cathode and backscattered with energies on the order of the discharge voltage. The bombardment of the growing film by these energetic species is referred to as atomic peening. The energy of both the sputtered species and the reflected neutrals will be reduced at high sputtering pressures because of collisions with other gas atoms occurring between the cathode and the substrate. The peening effect causes atoms to be incorporated into the growing film with a density higher than would be obtained otherwise. Similar bombardment occurs in films deposited using reactive low-voltage ion plating (RLVIP) or IAD.

In the models for intrinsic stress discussed previously, it is predicted that intrinsic stresses should not occur if substrate temperatures during deposition are greater than about $T_m/3$, where T_m is the melting temperature of the material.[2] This has been demonstrated for dual ion-beam-sputtered aluminum nitride (AlN) and evaporated metal films of Al, Ni, and Cu.[19, 20] The lack of stress occurs because the adatoms condensing on the film's surface have sufficient time and mobility to form an ordered stress-free film. In contrast, adatoms in films deposited at lower temperatures are virtually frozen in place. Because of this result, it is often thought that film stress can be reduced by depositing films at high temperatures. However, the total stress does not always decrease because the thermal stress component may increase more than the intrinsic stress component decreases.

6.3 Techniques for Modifying or Controlling Film Stress

Films with a low total stress are required for special coating applications where the stress of a coating would otherwise cause the film to fail mechanically. These applications usually involve thick multilayer coatings since the mechanical forces that cause a film to crack or delaminate scale with thickness. Examples of these coating applications include broadband reflective blocking coatings, infrared coatings, and rugate filters. Low-stress coatings are also required in cases where the coating stress would otherwise distort the optic by an unacceptable amount. Coatings with a low net stress can be produced by using materials with opposing stresses and designing the coating layer thicknesses so that the product of the stress and the thickness of all layers with a tensile stress equals that of all layers with a compressive stress. For example, consider an infrared blocking coating made up of several quarterwave stack reflectors at various wavelengths. If the low-index material has a tensile stress and the high-index material has a compressive stress of a greater magnitude, then the net stress of the entire coating will be compressive. The net stress of the entire coating can be reduced by making the low-index layers thicker and the high-index layers thinner. As long as the thickness of each pair of low- and high-index layers adds up to a halfwave optical thickness, the spectral performance of the multilayer reflector will stay about the same, although the width of the reflectance band and the peak reflectance will be reduced slightly. This approach limits the flexibility of the design and the material selection but it is easy to implement. Another approach is to reduce the stress of the individual layers of the coating. Several techniques have been developed for modifying or reducing the stress in optical coatings.

The most commonly used technique for modifying intrinsic stress is the variation of deposition parameters such as substrate temperature, deposition rate, and partial pressure of reactive gas. A variation in the deposition parameters affects the mobility of the deposited atoms, which affects the microstructure and the intrinsic stress of a film.[2] The fact that a change in deposition parameters can affect many of the film's other properties must be taken into consideration when modifying the deposition parameters to optimize a film's intrinsic stress. It may be that the parameters which produce the desired stress also result in a film with poor stoichiometry and increased absorption.

Effect of Deposition Parameters

Studies have shown that high substrate temperatures and low deposition pressures cause the intrinsic stress of films to be less tensile or more compressive. These deposition conditions increase the mobility of the deposited atoms, which results in films with a higher density and a more ordered structure.[16] Similarly, the intrinsic stress of films can be made less compressive or more tensile by using low substrate temperatures. This type of behavior was seen in a recent study investigating the deposition of ZnS films at substrate temperatures of −110 °C. Films of ZnS normally have compressive stresses when deposited at ambient or elevated temperatures.

In this study it was found that ZnS films deposited onto liquid-nitrogen-cooled substrates exhibited a tensile stress.[21]

It has been shown that the compressive stress of ZnSe films can be reduced by depositing the films at a high background pressure.[22] Films were deposited using electron-beam evaporation onto fused silica substrates heated to 150 °C. The effect of the partial pressure of O_2 on the total stress of ZnSe films is illustrated in Figure 6.3*a*. The film deposited without adding O_2 to the chamber had a high compressive stress, which is typical for ZnSe. For films deposited with a partial pressure of O_2, the stress decreased almost linearly as a function of pressure, and tensile stress values were obtained above an O_2 partial pressure of 1.3×10^{-4} torr. Films were also made using high partial pressures of Ar and N_2 to determine if the change in stress was

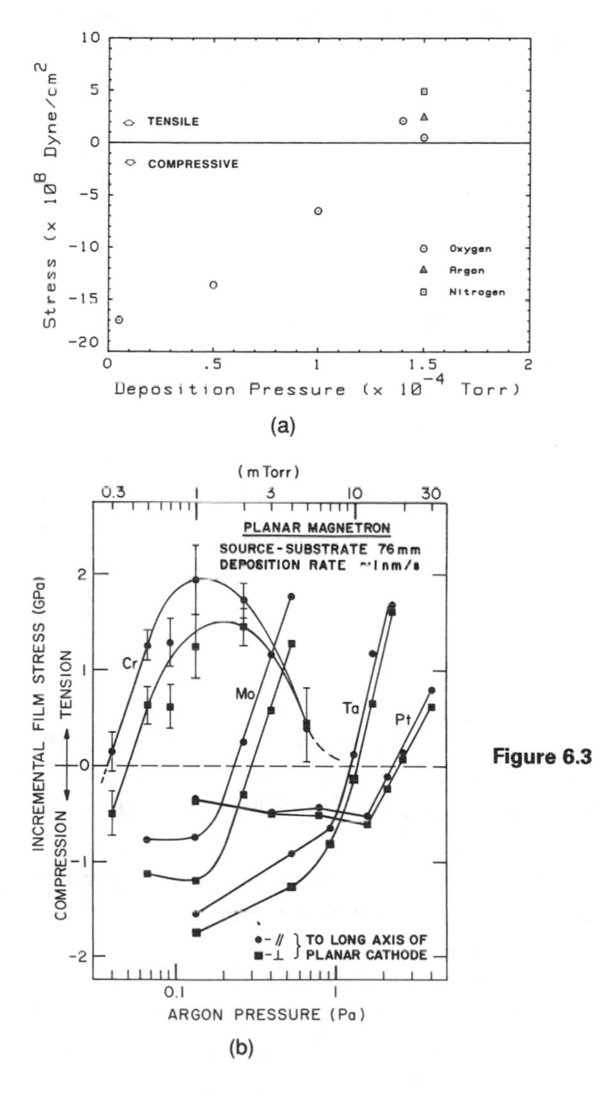

Figure 6.3 **Effect of deposition pressure on film stress: (*a*) Total stress of evaporated ZnSe films deposited onto fused silica substrates at various deposition pressures. O_2, Ar, and N_2 were used for backfill gasses.[22] (*b*) Film stress versus working pressure for Cr, Mo, Ta, and Pt metal films sputtered from a dc-planar-magnetron source onto glass at 1 nm/s to thicknesses of 50–350 nm.[25]**

caused by the formation of zinc oxide. These gases had effects similar to those of oxygen, indicating that the change in stress was caused by a decrease in the mobility of the deposited atoms due to gas scattering. This technique does show that ZnSe films with low stress can be obtained.

A dependence of the intrinsic stress of ZnSe films on the deposition rate has also been demonstrated.[14] In situ measurements have shown that the intrinsic stress of ZnSe films deposited at 150 °C could be reduced by a factor of two by increasing the deposition rate from 4 to 20 Å/s. It should be noted that the dependence of film stress on deposition parameters is highly material-dependent. For example, in the same study it was shown that neither deposition rate nor substrate temperature had any significant effect on the intrinsic stress of ThF_4 films.[14]

Deposition parameters can also affect the intrinsic stress of sputtered films. These deposition parameters include the pressure at which the films are deposited, the discharge or beam voltage, the geometry of the target in relation to the substrate, and the mass of the sputtering gas.[23] In reactive sputtering, the partial pressures of the reactive and the sputtering gases also affect the intrinsic stress.[24] Essentially, all these parameters affect either the energy or the flux of sputtered species and reflected neutrals that strike the substrate. Decreasing the energy or flux of these species tends to decrease the compressive stress or change the stress to tensile. Increasing their energy or flux tends to increase the compressive stress.

The effect of deposition pressure on the stress of sputtered films is illustrated in Figure 6.3*b*. In this example, films of chromium, molybdenum, tantalum, and platinum were deposited over a range of pressures using planar magnetron sputtering.[25] All the materials change from a tensile stress to a compressive stress as the pressure is reduced. At lower deposition pressures, the mean free path of the energetic species increases and these species suffer fewer collisions as they travel between the target and the substrate. As a result, the species have higher energy when they strike the growing film, and they drive the film to a compressive stress state. It is also noted that the stress changes rapidly as a function of pressure in the region of zero stress. This makes it difficult to produce low-stress films by controlling the pressure. The pressure at which the film changes from tensile to compressive is unique for each material.

The ratio of the mass of the sputtering gas to the target material can have a significant effect on the energy of the reflected neutrals, which affects the intrinsic stress of sputtered films. This effect has been demonstrated in the ion-beam sputtering of MgF_2 films.[3] The highly compressive stress of these films was reduced by one-half by using Xe as the sputtering gas instead of Ar.

In the reactive sputtering of dielectric materials the effects of deposition parameters on stress are highly material-dependent, and because the parameters in this process are so interdependent it is difficult to predict how a single process parameter affects the stress. For example, the compressive stress of aluminum oxide (Al_2O_3) films deposited from an Al target using reactive-dc-magnetron sputtering decreases as the discharge voltage increases, whereas films of SiO_2 deposited by the same process show no dependence of stress on discharge voltage. In addition, the stress of the Al_2O_3 films was lower for films deposited using higher partial pressures of Ar, but SiO_2 films deposited using a lower partial pressure of Ar had a lower stress.[26]

Effect of Ion-Assisted Deposition

Another technique which can be used to modify film stress is IAD. In this technique, an ion source is used to direct ions at a growing film during deposition. The bombardment of the growing film by energetic species increases the mobility of the deposited atoms and tends to change the intrinsic film stress to a more compressive state. Since most evaporated films have a tensile stress, this technique can be used to change the film stress to near zero. The modification of the film stress by IAD is dependent on a number of factors, including the ratio of the number of ions to the number of adatoms arriving at the substrate, the energy of the ions, and the gas species used in the ion source. Substrate temperature and deposition pressure can also affect the intrinsic stress of films deposited using IAD. A detailed study of the effects of IAD on the intrinsic stress of Nb films has been conducted.[6] At high deposition temperatures, the IAD process reduced the tensile stress of the films and the degree of change was dependent on both the ion energy and the current density. This result is consistent with the results obtained by others. At low substrate temperatures, it was found that the tensile stress of the films increased when IAD was used. This was explained in terms of removal of impurities in the IAD films which caused the film stress to become more tensile. This is in agreement with techniques discussed below where impurities are added to films to reduce the tensile stress.

Effect of Impurities

Another technique for modifying or reducing stress in optical coatings is the addition of impurities to the film. In some cases, only small amounts of an impurity are necessary to reduce the stress. This has the advantage that the refractive index is not significantly altered from its original value. In other cases, large compositional variations are required to achieve the desired stress reduction. These films often have refractive index values which differ significantly from the indices of the starting materials. Mixed-composition films can be deposited by several methods. Evaporated films can be deposited by coevaporation from two separate sources, or a single source can be used with impurities added to the starting material. It has been shown that impurities such as CaF_2 and ZnF_2 in the starting material of MgF_2 reduce the tensile stress of this material from 47×10^8 dyne/cm^2 to 19×10^8 dyne/cm^2. The impurities tend to congregate at the grain boundaries during film growth, reducing the surface energy and the stress.[1] It has also been shown that 2-μm-thick films of CeF_3 can be produced without stress cracks if the starting material contains approximately 20% barium fluoride (BaF_2). Otherwise, the thickness of CeF_3 films must be kept less than approximately 0.3 μm to avoid fracturing due to film stress.[4]

Some early studies of mixing materials to reduce stress suggested that materials with opposing stresses should be mixed together to obtain films with a low total stress. One of these studies investigated coevaporation of ZnS, which has a compressive stress, and thorium oxyfluoride ($ThOF_2$), which has a tensile stress. Films with a total stress of less than 0.06×10^8 dyne/cm^2 were obtained using this technique.[27] Thin plastic pellicles were coated successfully using this process, whereas most other films have stresses large enough to rupture the thin membrane. Further studies have shown that it is not necessary that the two starting materials have opposing stresses in order for low stresses with mixed composition films to be achieved.[28] An example of this is illustrated in Figure 6.4*a*. This graph shows the variation in intrinsic stress as a function of composition for coevaporated films made up of Ge and CeF_3. Both of these materials have high tensile stresses when evaporated separately, but mixed-composition films with 25% to 60% Ge by weight have very low compressive stresses. For both the CeF_3 and the Ge, the stress decreases rapidly as impurities are added. These films were deposited using coevaporation from two separate sources, and the CeF_3 was deposited using laser evaporation. Similar results were obtained with mixed-composition films of MgF_2 and Ge. Analysis of the films showed that the mixed-composition films had increased grain sizes and attributed the reduction in tensile stress to a decrease in grain boundary energy due to a decrease in grain boundary volume.

The compressive stress of ion-beam-sputtered films can also be reduced by adding impurities.[29] This is illustrated in Figure 6.4*b*. In this example, mixed-composition films of ZrO_2 and SiO_2 were deposited by cosputtering material from separate targets. The stress of the pure ZrO_2 film is reduced by a factor of five with the addition of just 10% SiO_2. This is a small enough fraction that the stress is reduced without significantly affecting the refractive index of the film. The significant reduction in the stress of the pure ZrO_2 film was correlated with a change in the crystalline phase of the film from polycrystalline to amorphous. Relatively low stress values were obtained for compositions with 10–70% SiO_2. For compositions greater than 70% SiO_2, the stress increases linearly to the stress of a pure SiO_2 film. This result is very similar to that presented in Figure 6.4*a* except that the stress does not change from compressive to tensile. Both examples do have a region where the stress is low and nearly constant for a broad range of compositions.

Effect of Post Deposition Annealing

Film stress can also be modified by baking. A post-deposition baking process changes a film's stress to a more tensile state due to film shrinkage. Even after deposition, films are still in a relatively high state of disorder. Baking these films, even at low temperatures, causes them to change to a more ordered state, which reduces their volume.[30] This has been demonstrated in both sputtered and evaporated films of optical materials.[5, 16] In evaporated coatings, this can pose serious problems for coatings which require several separate deposition and heating steps, since most evaporated coatings nominally have a net tensile stress. This stress can

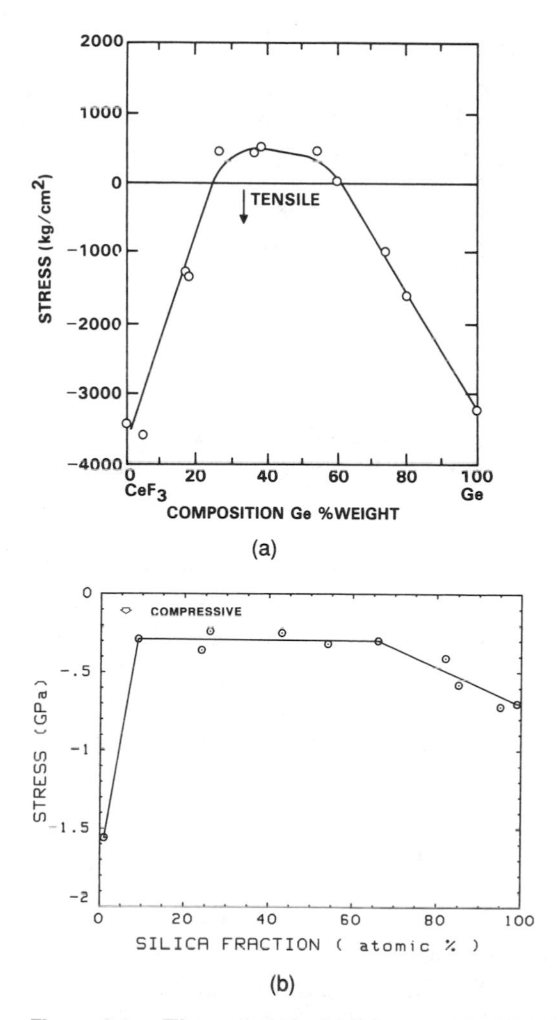

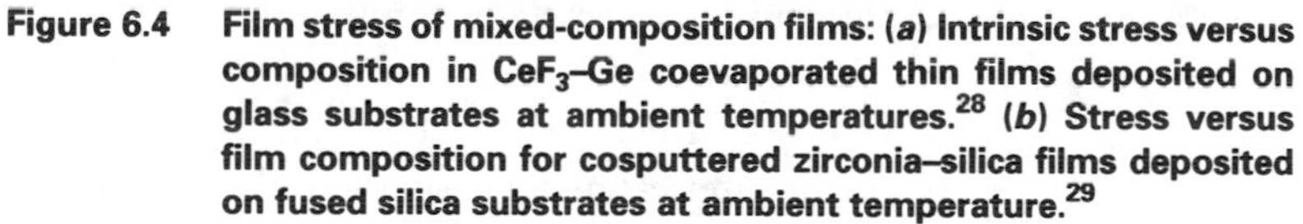

Figure 6.4 **Film stress of mixed-composition films: (*a*) Intrinsic stress versus composition in CeF_3–Ge coevaporated thin films deposited on glass substrates at ambient temperatures.[28] (*b*) Stress versus film composition for cosputtered zirconia–silica films deposited on fused silica substrates at ambient temperature.[29]**

increase to the point of failure when subjected to additional heatings. Baking the optical coatings also causes a change in the refractive index of the film, and, if the annealing temperature of a material is exceeded, the film's crystalline phase will change.

An advantage of baking is a reduction of the compressive stress in sputtered films.[5] An example of the change in stress of sputtered films due to a post-deposition baking step is illustrated in Figure 6.5. In this example, ion-beam-sputtered films of ZrO_2 and SiO_2 and mixed-composition films of ZrO_2–SiO_2 were subjected to a post-deposition bake at 300 °C in air for 3 h. The baking process

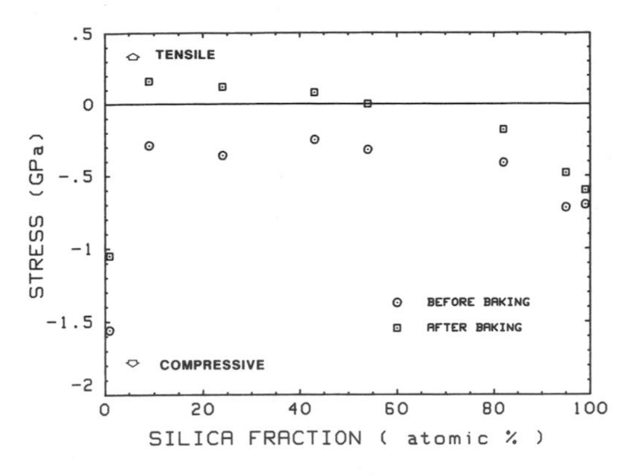

Figure 6.5 **Stress versus film composition for cosputtered zirconia–silica films, before and after baking at 300 °C for 3 h.**[5]

modified the stress of all the films, including the pure ZrO_2 and the pure SiO_2 films. For some compositions, the stress actually changed from compressive to tensile. The films with a higher ZrO_2 fraction changed by a greater amount, and it was found that the amount of change varied linearly with composition. Note that the pure ZrO_2 film and the film with a 90% ZrO_2 fraction initially have stress values that differ by a factor of five, yet both films change in stress by nearly the same amount due to baking.

6.4 Stress Measurement Techniques

Substrate Deformation

The most common methods for measuring film stress are based on the deformation of thin substrates due to the stress of a film. One method involves the use of flat circular disks which have a high aspect ratio, so they deform relatively easily.[7] This technique is easy to carry out, and substrates which are flat to approximately one wave can be easily obtained in a variety of thicknesses. The deformation of the substrate is typically measured using an interferometer[14] or a simple geometrical optic setup.[31] In either case the sag of the optic, or the deflection y of the center of the part relative to the outer edge, must be determined and then the stress σ can be calculated using the equation[32]

$$\sigma = \frac{yE_s(t_s)^2}{r^2 3(1 - v_s)\, t_f}$$

In this equation, E_s is Young's modulus for the substrate, v_s is Poisson's ratio for the substrate, t_s is the substrate thickness, r is the radius of the coated area being measured, and t_f is the film thickness. Since the sensitivity of this measurement is dependent on the aspect ratio of the substrate, very thin substrates 0.25–0.5 mm in

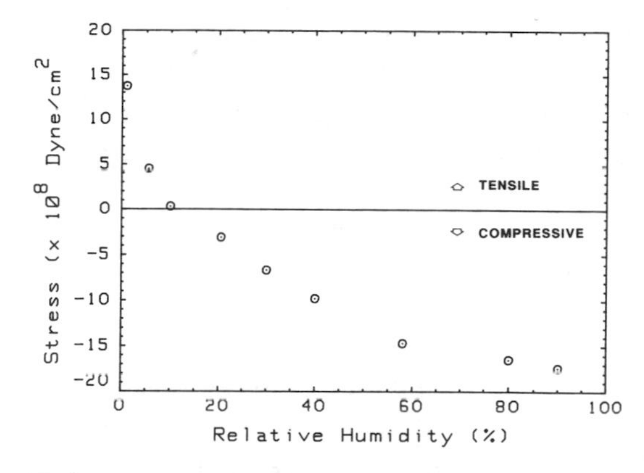

Figure 6.6 **Total stress versus relative humidity for an evaporated TiO_2/SiO_2 multilayer coating.[33]**

thickness are usually used. In addition, the deflection of the substrate should be measured prior to coating to ensure that an accurate change in deflection is obtained.

A relatively simple geometrical optic setup has been used to measure the deformation of coated disks in an environmental chamber to determine the effects of relative humidity and vacuum on multilayer coatings.[33] In this setup, a collimated beam of light is reflected from a stress witness and brought to a focus using a focusing lens. By measuring the focal length of the system and knowing the focal length of the focusing lens, one can determine the deformation of the stress witness. This system is easy to integrate into an environmental chamber and provides accurate measurements. An example of a film's dependence on relative humidity measured using the system described above is shown in Figure 6.6. This film is an electron-beam-evaporated multilayer coating deposited at a substrate temperature of 200 °C. The coating materials are TiO_2/SiO_2 and the film has a physical thickness of 7 μm. This result shows that the stress varies by 30×10^8 dyne/cm^2 as the relative humidity changes. A large part of the change occurs at relative humidities between 0% and 10%. In this region, the stress changes from a tensile stress of 13.5×10^8 dyne/cm^2 to almost zero. For the relative humidity range of 10–40% the change in stress is linear as a function of relative humidity. At higher relative humidity values, the rate of change in stress as a function of humidity decreases. This example shows that caution should be taken in regard to the environmental conditions in which ex situ stress measurements are made. For example, in a stress measurement being used to determine whether the coating stress would distort a production substrate by an unacceptable amount, it would be necessary to make the measurements under the same humidity conditions in which the production component would be used. Otherwise the stress-induced distortion could be severely under- or overestimated. It has been shown that films deposited by advanced deposition processes such as ion-beam sputtering and magnetron sputtering do not show a dependence of film stress on relative humidity.

Another measurement method which is based on the deformation of thin substrates is the cantilevered bending beam technique.[7] This method, used primarily for in situ stress measurements, can give very accurate measurements of film stress. In situ measurement of film stress is the only way to measure the intrinsic stress of a film accurately since it eliminates the thermal stress component and effects due to adsorption of moisture from the atmosphere. This approach enables one to determine the effects of deposition parameters and film thickness on the intrinsic stress of films. In situ measurements of film stress are difficult to perform since the apparatus must work in a vacuum chamber and must be operated remotely from outside the chamber. Other problems, including vibrations from pumps and operation at elevated temperatures, also make in situ measurements difficult. Some of the early in situ measurements of film stress were performed at ambient temperatures because of the difficulties associated with making these measurements at elevated temperatures.[34] However, film properties, including intrinsic stress, are strongly dependent on substrate temperature, a fact which must be taken into account in evaluating these results.

In the cantilevered beam apparatus, a thin glass plate is clamped at one end and the other end is freestanding. The deflection y of the free end of the beam can be monitored in a number of ways, including observing with a microscope, measuring the deflection of a reflected laser beam, or making a differential capacitance measurement in which the free end of the beam is part of the capacitor. In all these techniques, the stress of the film is determined using the equation

$$\sigma = \frac{E_s h^2 y}{3l^2(1 - v_s)t_f}$$

where l is the length and h is the thickness of the beam.

An extensive study of the intrinsic stress of a number of optical coating materials was conducted by measuring the distortion of a thin strip using an interferometer.[34] All the films were deposited at ambient temperatures, and a number of visible and infrared coating materials were investigated. One interesting result from this study was the measurement of the stress of a multilayer coating. The stress of this coating as a function of thickness is shown in Figure 6.7. The low-index material in this coating was $ThOF_2$, and this film has a tensile intrinsic stress. The high-index material was ZnS, which has a compressive intrinsic stress. The variation in the substrate curvature decreased with each successive layer and tended toward an average value indicating a net compressive stress. Several models have been developed for predicting the stress in multilayer coatings based on the measured stress of the individual layers. One of these models uses a very simple calculation, but poor agreement between the predicted and the measured stress was obtained using this technique.[34] A more rigorous analysis has been developed, but the calculations in this technique are complex and the accuracy has not been demonstrated.[14]

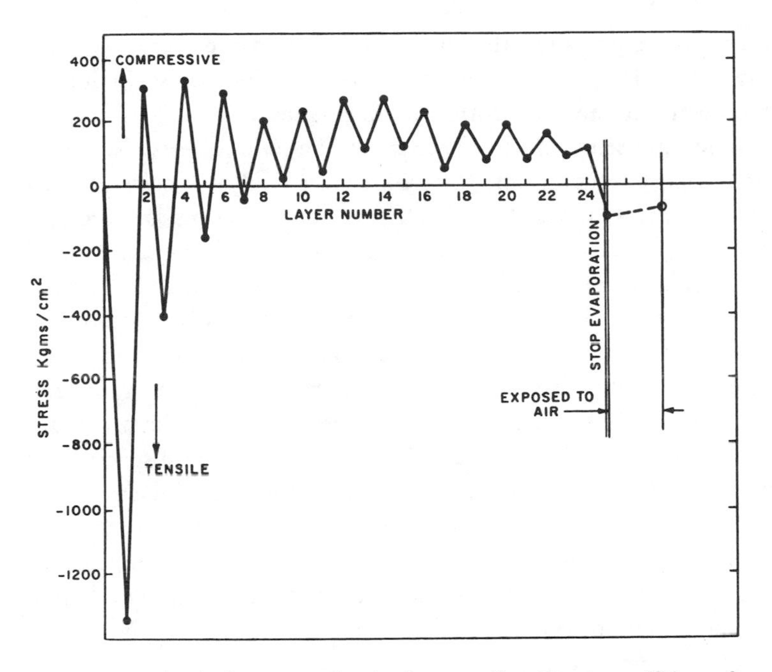

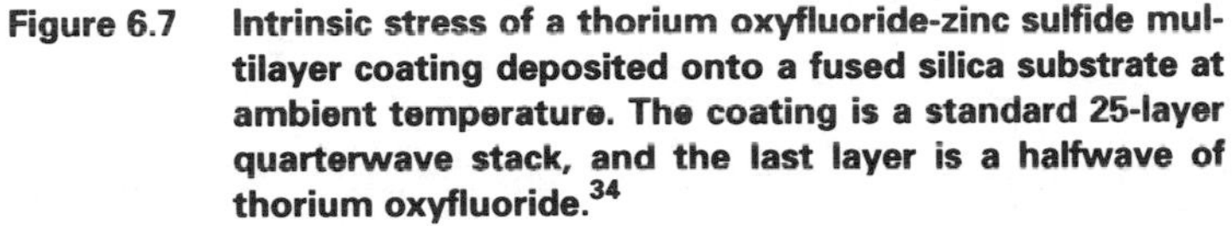
Figure 6.7 **Intrinsic stress of a thorium oxyfluoride-zinc sulfide multilayer coating deposited onto a fused silica substrate at ambient temperature. The coating is a standard 25-layer quarterwave stack, and the last layer is a halfwave of thorium oxyfluoride.[34]**

X-Ray Diffraction (XRD)

A commonly used technique which determines the crystalline phase of a material or film is XRD. This technique can also be used to measure the stress in a thin film, although it cannot be used to measure the stress of amorphous materials.[8] Basically, XRD determines the spacing between parallel planes of atoms in a crystal lattice from the angular location of peaks in an X-ray diffraction pattern. In thin films, the lattice is often in a stressed state and the spacings are compressed or enlarged in comparison to the spacings observed in unstressed solid crystals of the same material. The film stress can then be determined from the difference between the measured lattice spacing of the film and the lattice spacing of an unstressed crystal of the same material.

In XRD measurements of thin films, the diffraction peaks are often quite broad since the films have lattice spacings that vary slightly from grain to grain and because the grain sizes are small. In addition, the signals obtained from films are usually small because the path length is small. This makes it difficult to measure the location of an XRD peak accurately, which limits the accuracy of the stress measurement. An advantage of this technique is that a special kind of substrate is not required for the measurement since the lattice strain is measured directly. Since XRD measurements are made ex situ, they measure the total film stress, including thermal stress components and any adsorption-induced stresses.

The XRD technique was used to measure the stress of a series of mixed-composition films deposited using coevaporation.[35] In some cases ZrO_2 was coevaporated with SiO_2 and in other cases with MgO. A pure ZrO_2 film was also produced and analyzed. The crystalline phase of all the films was tetragonal, and the *d* spacing of the (101) plane was calculated using the location of the corresponding XRD peaks. In order for the stress to be calculated, the Young's modulus and Poisson's ratio for the film are required. In these calculations, values of 1.3×10^{12} dyne/cm^2 and 0.23 were used. Since the tetragonal phase of ZrO_2 is not stable at low temperatures, a value for the *d* spacing of the (101) plane was extrapolated from high-temperature data, and the value obtained was $d_{(101)} = 0.2957$ nm. The measured *d* spacing for the pure ZrO_2 film was $d_{(101)} = 0.2955$ nm, and the calculated film stress was 45.0×10^8 dyne/cm^2 tensile. This shows that a very small change in the spacing of the crystal planes (0.0002 nm) produces a measurable film stress. A number of other samples were measured in the study; these showed that the coevaporated films had tensile stresses higher than the pure ZrO_2 film. The film with the highest stress had a composition of 59% ZrO_2 and 41% MgO.

Raman Spectroscopy

Raman spectroscopy has been used extensively for probing the structure of coatings, and recently its use for measuring film stress has been demonstrated.[9] This is a nondestructive technique which has the ability to probe multiple film properties such as stoichiometry, grain orientation, impurity content, and film stress. The technique can be used for in situ measurements to monitor a film's properties during growth and it can be used to analyze both single- and multilayer films. Focusing the laser probe light to a diffraction-limited spot size makes it possible to obtain spatially resolved Raman scattering (SRRS) from micron-sized regions in a film. Raman spectroscopy is performed by illuminating a sample with a continuous wave (CW) or pulsed laser probe, and the Raman-scattered light is collected and directed to a monochromator. This is an inelastic scattering process so the scattered photon may have less energy or more energy than the incident photon, depending upon the vibrational state of the molecule with which the incident photon interacts. Raman scattering is a relatively weak effect, and studying thin films compounds the problem since the interaction volume is so small. The substrate also contributes its own Raman spectrum, which may be difficult to distinguish from the spectrum of the film. A number of techniques have been developed for minimizing the contribution from the substrate, including waveguiding the probe beam in the film.[36] Techniques have also been developed for enhancing the Raman signal in multilayer films by designing the coatings to enhance the electric field of the incident laser light in the layers being probed.[37]

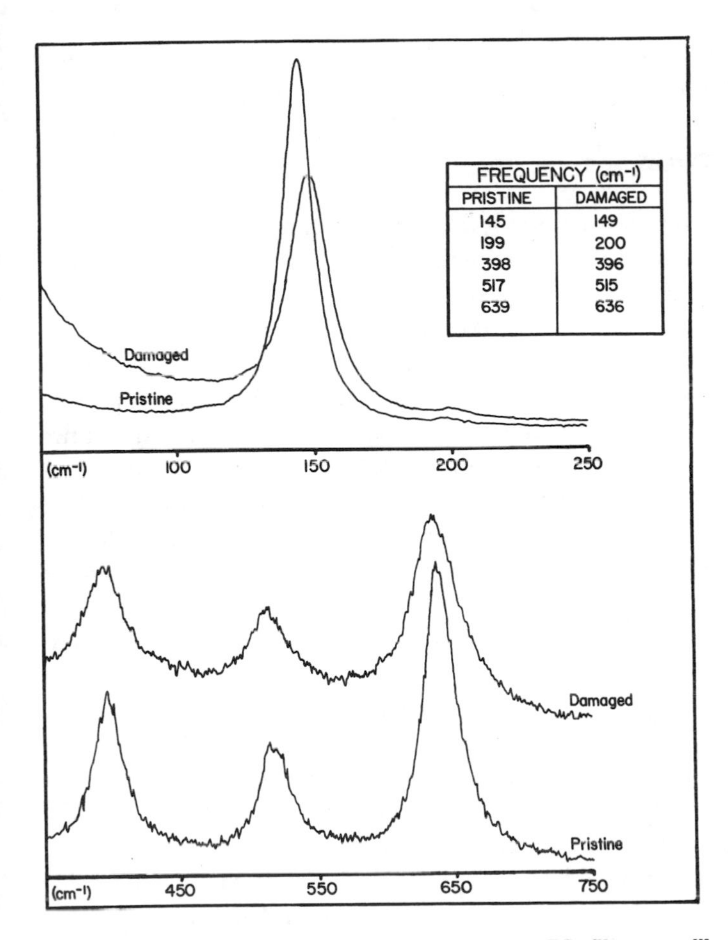

Figure 6.8 **Raman spectra of a 0.9-μm anatase TiO_2 film on a silica substrate before and after exposure to a single 532-nm high-energy pulse.**[9]

Stress in thin films causes the peaks of the Raman spectrum to shift in frequency from those of an unstressed sample.[9] An example of a Raman spectrum shifted in frequency due to stress is shown in Figure 6.8. This spectrum is of a laser-damaged region of a reactively sputtered anatase TiO_2 film. The film was irradiated with a high-energy 532-nm pulsed laser, and the measurement was made with a Raman microprobe. This sample was irradiated at a pulse energy much lower than the damage threshold, yet frequency shifts indicate a compressive stress was induced in the coating. The measured frequency shifts correspond to a bulk compressional increase of approximately 100×10^8 dynes/cm^2. At higher energies, the films recrystallize to the rutile phase and damage catastrophically. Analysis of a catastrophically damaged coating using the same technique showed that the stress-induced frequency shift measured near the edge of a damage site decreased rapidly 100 μm away from the edge.[9] This example illustrates the unique capabilities and versatility of Raman spectroscopy in the measurement of stress, especially when used in a microprobe configuration.

6.5 Future Directions

There is still a great deal to be learned about film stress in thin film optical coatings. With film stress measurements there is a distinct lack of data on stress values of optical coating materials. A few materials such as MgF_2 have been studied thoroughly, but little has been published on most optical coating materials. There is also a limited amount of information published on environmental effects on stress in optical coatings. A recent study utilized a unique cantilevered-beam apparatus which could be used in a coating chamber to measure stress and then the entire apparatus could be transferred to an environmental chamber for further measurements.[16] This type of stress measurement apparatus is very versatile and could be used for measuring film stress under various environmental conditions, including extreme environments such as those in deep space. Such an apparatus could also be used to measure aging effects in film stress.

In order to obtain a fundamental understanding of the relation of film stress to other film properties during film growth, it is necessary for analysts to use in situ measurements. In situ Raman spectroscopy has been proposed for this type of measurement since it can measure the microstructure of a film in addition to its stress. Another possible measurement technique is in situ ellipsometry, which is becoming increasingly valuable for monitoring the growth of thin films.[38] In addition, its ability to determine the microstructure of a film during growth has been demonstrated. This technique could be used in combination with a cantilevered-beam apparatus to measure simultaneously a film's stress and its microstructure during growth. Such studies could help analysts develop a better understanding of the origins of stress in optical coating materials.

Some general rules, such as the observation that higher deposition pressures cause stress to tend towards a tensile state, have been developed for estimating the effects of deposition parameters on film stress. However, it would be desirable to have predictive theories for each class of materials and for the various deposition processes. A significant amount of work along these lines has been performed with thin films deposited by sputtering, although most of these studies involved metal films.[23] Studies of this type need to be conducted with the various classes of optical coating materials deposited by both sputtering and evaporation. The results of such studies would be very helpful in the production of optical coatings.

The design of optical coatings is driven by the spectral requirements, and it is anticipated that future applications will have more difficult spectral requirements. In order to achieve these requirements, it will be necessary for analysts to control the stress in materials so a broader base of materials will be available to choose from. It is anticipated that approaches such as the one utilized for the production of rugate filters will be a future direction for solving stress problems.[28] In this approach, materials were mixed together to produce new materials which had very low stresses. These mixed composition films were also dense so their stress was not dependent on the relative humidity.

In some cases, the role of stress is yet to be determined. For example, the effects of film stress on laser damage threshold are still uncertain. An early study showed a definite relation between film stress and laser damage threshold,[39] while a more recent study showed no correlation between these two parameters.[40] In both of these studies, many film properties besides stress were also affected. This is because most film properties are interdependent. It may have been film properties other than stress that determined the laser damage threshold in both studies. In order to determine the role of stress in laser damage accurately, analysts must design an experiment in which stress is an independent variable. A possible approach may involve using a process with many parameters, like IAD or RLVIP, which would allow analysts to vary the stress while maintaining other film properties at fixed values.

Acknowledgment

I would like to thank C.K. Carniglia for his help in writing this article. I would also like to thank H. Windischmann for his helpful suggestions.

References

1 H. K. Pulker and J. Mäser. *Thin Solid Films.* **59**, 65, 1979.

2 F. M. D'Heurle and J. M. E. Harper. *Thin Solid Films.* **171**, 81, 1989.

3 T. H. Allen, J. P. Lehan, and L. C. McIntyre, Jr. *Proceedings.* Vol. 1323. SPIE, 1990, p. 277.

4 S. F. Pellicori. *Thin Solid Films.* **113**, 287, 1984.

5 B. J. Pond, J. I. DeBar, C. K. Carniglia, and T. Raj. NIST (U.S.) Spec. Publ. 775. 1988, p. 311.

6 J. J. Cuomo, J. M. E. Harper, C. R. Guarnieri, D. S. Yee, L. J. Attanasio, J. Angilello, C. T. Wu, and R. H. Hammond. *J. Vac. Sci. Technol.* **A20**, 349, 1982.

7 D. S. Campbell. In *Handbook of Thin Film Technology.* (L. I. Maissel and R. Glang, Eds.) McGraw-Hill, New York, 1970, chapt. 12.

8 B. D. Culity. *Elements of X-ray Diffraction.* Addison-Wesley, Reading, MA, 1978.

9 G. J. Exarhos. *J. Vac. Sci. Technol.* **A4**, 2962, 1986.

10 K. H. Guenther. *Appl. Opt.* **23**, 3612, 1984.

11 R. R. Zito. *Thin Solid Films.* **87**, 87, 1982.

12 K. L. Chopra. *Thin Film Phenomena.* McGraw-Hill, New York, 1969.

13 M. L. Scott. NBS (U.S.) Spec. Publ. 688. 1983, p. 329.

14 A. M. Ledger and R. C. Bastien. Technical Report. Contract DAAA25-76-C-0410 (DARPA). Perkin-Elmer Corp., Norwalk, CT, 1977.

15 R. W. Hoffman. *Thin Solid Films.* **34**, 185, 1976.

16 H. O. Sankur and W. Gunning. *J. Appl. Phys.* **66**, 807, 1989.

17 G. K. Wehner and G. S. Anderson. In *Handbook of Thin Film Technology.* (L. I. Maissel and R. Glang, Eds.) McGraw-Hill, New York, 1970, Chapt. 3.

18 J. A. Thornton and D. W. Hoffman. *Thin Solid Films.* **171**, 5, 1989.

19 H. Windischmann. *Thin Solid Films.* **154**, 159, 1987.

20 E. Klokholm. *J. Vac. Sci. Technol.* **6**, 138, 1969.

21 M. D. Himel, J. A. Ruffner, and U. J. Gibson. *Appl. Opt.* **27**, 2810, 1988.

22 J. Sobczak, B. J. Pond, C. K. Carniglia, R. A. Schmell, and M. F. Dafoe. *OSA Annual Meeting Technical Digest MNN2.* Orlando, FL, 1989.

23 H. Windischmann. *J. Vac. Sci. Technol.* **A9**, 2431, 1991.

24 G. Este and W. D. Westwood. *J. Vac. Sci. Technol.* **A5**, 1892, 1987.

25 D. W. Hoffman and J. A. Thornton. *J. Vac. Sci. Technol.* **A20**, 355, 1982.

26 B. J. Pond, T. C. Du, J. Sobczak, C. K. Curniglia, and F. L. Williams. *Proceedings.* Vol. 1624. SPIE, 1991, p. 174.

27 R. J. Scheuerman. *J. Vac. Sci. Technol.* **7**, 143, 1970.

28 H. Sankur, W. J. Gunning, and J. F. DeNatale. *Appl. Opt.* **27**, 1564, 1988.

29 B. J. Pond, J. I. DeBar, C. K. Carniglia, and T. Raj. *Appl. Opt.* **28**, 2800, 1989.

30 H. Oikawa and Y. Nakajima. *J. Vac. Sci. Technol.* **14**, 1153, 1977.

31 S. M. Rossnagel, P. Gilstrap, and R. Rujkorakarn. *J. Vac Sci. Technol.* **21**, 1045, 1982.

32 A. Brenner and S. Senderoff. *J. Res NBS.* **42**, 105, 1949.

33 B. J. Pond, H. Nusbaum, T. C. Du, J. Sobczak, R. A. Schmell, C. K. Carniglia, R. O. Petty, J. Johnston, and G. B. Charlton. *Technical Digest, Topical Meeting on Optical Interference Coatings.* Optical Society of America, Washington, DC, 1992.

34 A. E. Ennos. *Appl. Opt.* **5**, 51, 1966.

35 E. N. Farabaugh and D. M. Sanders. *J. Vac. Sci. Technol.* **A1**, 356, 1983.

36 A. F. Stewart, D. R. Tallant, and K. L. Higgins. NBS (U.S.) Spec. Publ. 746. 1985, p. 362.

37 R. A. Craig, G. J. Exarhos, W. T. Pawlewicz, and R. E. Willford. *Appl. Opt.* **26**, 4193, 1987.

38 J. K. Moyle, W. J. Gunning III, and W. H. Southwell. *Proceedings.* Vol. 821. SPIE, 1987, p. 157.

39 R. R. Austin, R. Michaud, A. H. Guenther, and J. Putman. *Appl. Opt.* **12**, 665, 1973.

40 F. Rainer, W. H. Lowdermilk, D. Milam, C. K. Carniglia, T. Tuttle Hart, and T. L. Lichtenstein. *Appl. Opt.* **24**, 496, 1985.

7

Surface Modification of Optical Materials

CARL J. McHARGUE

Contents

7.1 Introduction

The use of energetic ion beams to modify the near-surface region of materials has been investigated extensively in recent years. It is an accepted industrial procedure for doping semiconducting materials and for increasing the wear, erosion, and oxidation resistance of metallic alloys. Use of this technique for modifying the structure and properties of the surfaces of structural ceramics and of optical materials is under active study.

The process consists of bombarding the surface of a material in a vacuum chamber with an electrostatically accelerated beam of ions. Because of the kinetic energy, the ions become embedded to a depth controlled by the incident ion energy for a given ion–target material combination. This range is generally less than a micrometer. The ion comes to rest by dissipating its kinetic energy in elastic collisions that displace target atoms (ions) from their normal lattice sites and by inelastic (electronic) processes that may ionize target atoms. Large numbers of defects are produced before the ion comes to rest as an impurity, dopant, or alloying element.

The nature of the ion implantation process permits the introduction of any element into the near-surface region of any solid in a controlled and reproducible manner that is independent of most equilibrium constraints. Since the process is nonequilibrium in nature, compositions and structures unattainable by conventional methods may be produced.

Many features of the damage microstructure in metals are well understood because of the large body of literature generated by studies of radiation damage induced by neutron bombardment in nuclear reactors. In many cases, there are reasonably accurate theoretical models to describe the evolution of the microstructure as a function of point defect concentration.

Implantation- or radiation-induced damage in insulators is much more complex and less studied. These materials are generally compounds composed of at least two sublattices (in the crystalline case) that have different atomic masses. A different energy to displace each type of atom (ion) from its lattice site may be required. The types of defects produced are strongly influenced by requirements of local charge neutrality, the local stoichiometry, and the nature of the chemical bond in the particular lattice. In addition, inelastic (electronic or ionizing) processes may produce lattice defects, whereas in metals such effects are unimportant.

This chapter first treats some fundamentals of ion–solid interactions in terms of the energy-loss mechanisms, range, and cascade formation. Relationships among defect formation, rearrangements, and the resulting microstructure are then discussed. Alteration of optical properties is given for a few materials. Detailed discussions of ion–solid interactions,[1] point defects in the materials of interest,[2] and ion implantation in these materials[3–5] are found in the referenced works.

7.2 Fundamental Processes

Ion–Solid Interactions

An energetic ion striking a target loses energy continuously by collisions with the nuclei and electrons of the target atoms until it reaches thermal energies. The time required to come to rest is of the order of 10^{-14} s. The major mechanisms of energy loss are (1) direct collisions with the screened nucleus of a target atom and (2) excitation of electrons bound to such atoms. Each process is energy-dependent and makes different contributions to the energy loss along the ion's path. To a reasonable approximation, these are independent processes so that the linear rate of energy loss is given by the sum of two contributions:

$$\frac{dE}{dx} = \left(\frac{dE}{dx}\right)_{\text{nuclear}} + \left(\frac{dE}{dx}\right)_{\text{electronic}} \tag{7.1}$$

This differential function is known as the stopping power and is often represented by S.

Nuclear stopping usually dominates at lower energies, and electronic stopping dominates at higher energies. Both processes make significant contributions in the region typically of interest for ion implantation. The nuclear stopping is important in determining ion ranges, displacement damage, and sputtering. The electronic stopping is important for excitation phenomena that result in secondary electron emission and that can produce defects in insulators.

The stopping power can be used to calculate the total path length of an ion and the distribution of radiation damage along the path or the distribution of the implanted ion. The total path length (or range) can be estimated by

$$R_T = \int_0^{E_0} - \frac{dE}{(dE/dx)_T} \tag{7.2}$$

where $(dE/dx)_T$ sums the nuclear and electronic contributions to the total stopping power and E_0 is the incident ion energy. Generally, the projected range relative to the incident direction is of more interest. It describes the spatial distribution of implanted ions relative to the surface. To a first approximation, the final ion distribution calculated from transport equations is given by a Gaussian function. The values for the range and its first two moments are given as a function of energy for a number of ion–target combinations of interest to us in Reference 3. The Gaussian distribution often underestimates the peak concentration on highly skewed distributions, that is, for higher energies and lighter ion masses.

Defect Production, Rearrangement, and Retention

The nuclear component of the energy loss is dissipated in elastic collisions which cause atom displacements for energy transfers greater than some threshold value. This threshold for displacement, E_d, lies in the range of 20–40 eV for most metallic materials and can range up to more than 100 eV in some compounds. The target atom initially struck by the incident ion may receive an energy that is considerably greater than E_d and thus can cause additional displacements along with the incident ion. A cascade of collisions and defect production ensues. This recoil atom also loses its energy by both nuclear (collisions) and electronic (ionizing, excitation) processes.

Cascades that involve a relatively small number of atoms in motion at a given moment, that is, isolated collision cascades or the low fluence limit, are termed linear cascades. Analytical techniques based on both transport equations and Monte Carlo techniques have been used to give the depth profile of deposited damage energy and the initial distribution of defects. The Monte Carlo method simulates individual collision cascades which are then averaged over several hundred such events. Both approaches give the initial distribution of defects.

The basic unit of radiation-induced defects in crystals is the Frenkel pair: one vacancy and one self-interstitial. In the low-density cascade regime, the number of defects produced can be related to the deposited damage energy, $S_D(x)$, by the modified Kinchin-Pease relationship as

$$dpa(x) = \frac{0.8\tau S_D(x)}{N2E_d} \tag{7.3}$$

where *dpa* indicates displacements per atom, τ is the fluence of bombarding ions, N is the number density of target atoms, E_d is the target displacement energy, and

$S_D(x)$ is the deposited damage energy at depth x. In the first part of the ion's path, the vacancy and its interstitial are separated by several interatomic distances. However, in the dense cascades and at end-of-range, there is a large concentration of both, consisting of a vacancy-rich "core" surrounded by a "shell" of interstitials.

The amount of "radiation damage" or defect production caused by these elastic collisions during implantation is substantial. For example, at the position of peak damage, the ratio of aluminum ions displaced to the number of target aluminum ions is 4.1 for implantation of 4×10^{16} Cr^+/cm^2 (150 keV) into Al_2O_3. Of course, only a small fraction of these vacant Al-sites survive the cool-down period after a collision cascade.

The energy partitioned to electronic excitation can also produce defects in some insulators. The depth profile for electronic energy losses (also called inelastic losses) can be calculated in the same manner as for nuclear processes. Since electronic stopping dominates at higher energies and nuclear stopping at lower energies, the deposited electronic energy is a maximum at the surface and falls off rapidly. The relative amount of energy deposited in the two processes varies strongly with incident energy and with both ion and target atom atomic numbers. Most of the energy is lost in electronic processes for very light ions.

There must be an efficient coupling of the electronic energy to the lattice for the production of atomic displacements to occur. Townsend and Agullo-Lopez[6] have listed four general requirements for the occurrence of such displacements: (1) The electronic excitation must be localized to one or a few lattice sites; (2) The local excited state must last long enough to couple into a mechanical response of the nuclear masses (i.e., comparable to phonon periods); (3) The excitation energy must be equal to or greater than the displacement energy for the atom in this excited state; and (4) The primary damage products (vacancies and interstitials) should be separated by several lattice sites in order that recombination be prevented.

These considerations eliminate this mechanism from metals, insulators, and narrow-gap semiconductors, leaving single-electron excitations in wide-gap semiconductors as likely candidates. The process is relatively well-understood for the alkali halides, A^+X^-, where it produces color centers. It is also the basis for the photographic process in silver halides. Although the mechanism is not understood, ionization-induced defects have been identified in SiO_2.[2]

Additionally, ionizing radiation may induce or enhance other reactions or processes in the solid state. These include recombination-assisted dislocation climb and recombination-assisted annihilation of deep traps in III–V compounds and the dissociation and luminescence quenching of hydrogen-containing centers in H-implanted SiC.

The microstructural changes are caused by the defects that survive the "cooling-down" period after the cascade, of the order of 10^{-10}. Many of the vacancy-interstitial pairs recombine and annihilate each other during this period. However, observations by electron microscopy indicate that defect clusters can be directly formed in the cascade. Interstitials can cluster into dislocation loops, while vacancies may form dislocation loops, stacking fault tetrahedra, or even three-dimensional cavities.

Changes in the properties of ion implanted regions result from (1) microstructural changes due to defect generation and rearrangements, (2) compositional changes, (3) radiation-induced phase changes, and (4) residual stresses. Implantation parameters that strongly affect the microstructural and phase changes are substrate temperature, ion fluence, and chemical nature of the implanted species. Ion beam energy generally affects only the distribution of the damage and implanted ions.

The effect of ion implantation on the specific defects generated and the changes in optical properties are discussed for selected materials. The discussion is not exhaustive but is intended to indicate the possibility of tailoring materials for specific applications.

7.3 Ion Implantation of Some Optical Materials

Glasses and Amorphous Silica

Defect generation leading to random atomic arrangements is not limited to crystalline materials. Amorphous (or glassy) materials contain various atomic packing densities and are composed of some type of fundamental structural unit, for example, SiO_4 tetrahedra in fused silica. Amorphous silica relaxes to a higher density during ion implantation due to both electronic excitation and nuclear (elastic collisions) damage processes. The defects produced by the two processes are different as shown by the annealing behavior of nitrogen implanted silica. All electronic damage can be removed by an anneal at 450 °C, leaving the elastic displacement damage (vacancies) behind.

Two types of point defects have been identified for high-purity amorphous silica. The E′-center is an electron trapped in a nonbonding sp^3 hybrid orbital on a silicon atom at the site of an oxygen vacancy. There is a local rearrangement and relaxation of the structure, that is, compaction or densification. The E′-center can be produced by both electronic (ionizing) processes and collisional displacements. A second type of defect, the B_2-center, is produced by collisional displacements only. It is described as an oxygen vacancy containing a trapped electron or electrons.

Figure 7.1 shows optical absorption data from Arnold[7] for hydrogen- and xenon-implanted Corning silica 7940. The H-implanted sample contains only the E'_1-center, whereas the Xe-implanted sample contains both E'_1- and B_2-centers. Essentially all (99.8%) of the energy (250 keV) of the hydrogen ion was deposited via electronic processes (inelastic interactions), confirming that E'_1-centers can be produced by such interactions. Most (88.7%) of the energy of the xenon ion was dissipated in nuclear events (elastic collisions) producing both E'_1- and B_2-centers.

The compaction of silica and simple silica glasses saturates at a value of about 2.8%.[8] This increase in density leads to a maximum increase in index of refraction (n) of 2–4%. The changes in volume, $\delta V/V$, and in index of refraction, $\delta n/n$, are plotted as functions of deposited nuclear (collisional) damage density in Figure

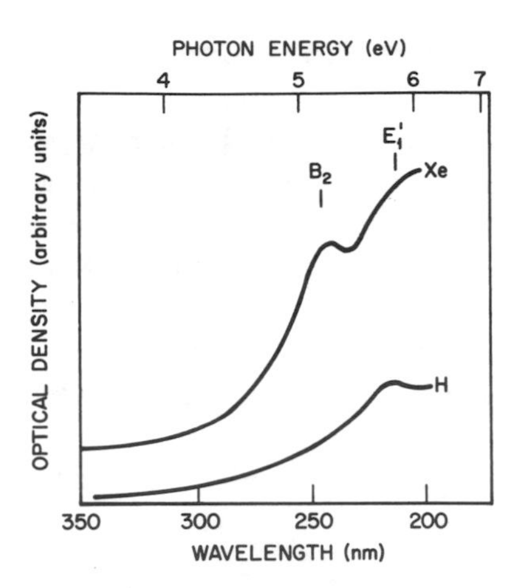

Figure 7.1 **Optical absorption of Corning silica 7940 implanted with Xe or H. Only the E_1' defect is produced by the mostly electronic damage of the hydrogen, whereas both E_1' and B_2 defects are produced by elastic damage of xenon. (From Arnold.[7])**

7.2. In this figure, data for $\delta n/n$ from Hines and Arndt[9] and those for $\delta V/V$ from EerNisse[8] are replotted to indicate the correspondence between them. The maximum volume decrease is 2.8%, and the index of refraction at the highest fluence is 1.475, $\lambda = 575$ nm. The volume change reaches a saturation value at a damage energy of about 1×10^{20} keV/cm^3 and the change in index at about 2×10^{20} keV/cm^3. Simple models relating δn to δV appear to be accurate to about the same degree.

The change in refractive index can be larger than that already indicated if the implanted ion produces "chemical" changes in addition to defect production. Webb and Townsend[10] noted ~6% increase in the case of nitrogen-implantation to a fluence of 2×10^{16} ions/cm^2. In this instance, the nitrogen was incorporated into the glass structure, possibly as silicon oxynitride or silicon nitride.

On the other hand, ion bombardment of soda-lime-silica glass *decreases* the index in the implanted layer, forming an antireflecting surface. This decrease is due to the depletion of sodium from the surface. Alkali depletion from alkali-containing glasses is a well-documented radiation effect, occurring for both electron and ion irradiation. The results of many studies favor a model of radiation-enhanced migration of the alkali ions within the implanted region and radiation-induced segregation to the surface.

The presence of E′- and B_2-centers gives optical absorption bands at 5.8 and 5.0 eV, respectively (Figure 7.1).

Luminescence centered about 2.7 eV is observed even during ion implantation of fused silica and is associated with the E′-centers. It has been extensively studied at the University of Sussex (Townsend and students).

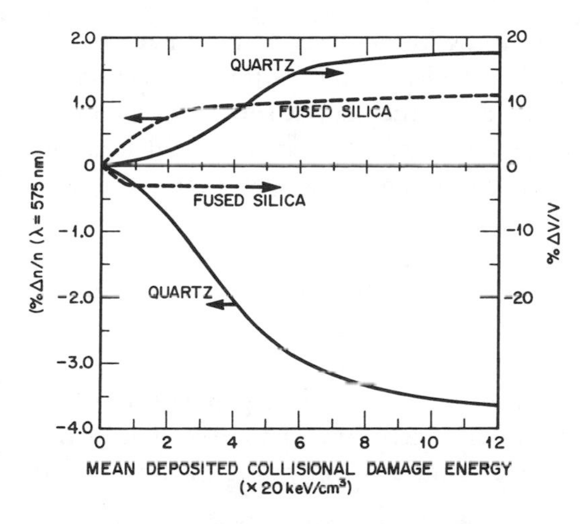

Figure 7.2 **Percentage change in volume and refractive index for ion implanted-quartz and amorphous silica normalized to collisional damage energy deposition. (Data from Hines and Arndt[9] and EerNisse.[8])**

α-Quartz (SiO_2)

The principal defects in pure crystalline silica (α-quartz) are also the E′- and B_2-centers. Two variants of the E′-type have been identified from optical absorption measurements. The E_1'-center (absorption band at 5.8 eV) has the hybrid orbital parallel to the "short" Si–O bond direction; the E_2'-center (absorption band at 5.3 eV) has the hybrid pointing parallel to the "long" Si–O bond direction.

The electronic (radiolysis) process that is active in producing the E_2'-center in amorphous silica at room temperature produces defects in quartz only below about 100 K. The vacancy–interstitial pairs so produced are so close together that they recombine (anneal) upon warming to room temperature.

Additional variations of these two types of centers have been detected by various spectroscopic techniques. They are often associated with the presence of hydrogen. There are also impurity-induced defects, the A-centers.

By far, the most important "defect" produced by the ion beam is a region of highly disordered material along the ion trajectory. These regions contain a random distribution of the host atoms and are characterized as "amorphous" from Rutherford backscattering spectroscopy (RBS) experiments.

Crystalline quartz expands upon irradiation and its refractive index decreases. These changes are shown in Figure 7.2 The volume expansion reaches a maximum of 19% at a mean collisional damage energy density of about 2×10^{21} keV/cm^3. The refractive index decreases from an unimplanted value of 1.545 (λ = 575 nm) to a saturation value of 1.482, a decrease of ~4%. This decrease in index is ascribed to the amorphization of the crystalline SiO_2 by ion irradiation.

Götz[3] has divided the damage production with ion fluence into three regimes. In the first stage (below ~1×10^{20} keV/cm^3), there is a slow increase in disorder (i.e., the number of Si ions that do not reside on their lattice sites) with increasing fluence. The damage at this stage can be described as isolated point defects and simple defect clusters. There is no detectable volume change, but there is a small change in index (~0.4%).

In the second stage (up to ~2.5×10^{20} keV/cm^3), damage accumulates rapidly to a completely random distribution, as indicated by RBS measurements. Götz suggests that each incident ion produces a microregion of amorphous material around its path. Both volume and index also change rapidly with fluence in this stage.

The third stage, saturation of damage, corresponds to overlapping of the amorphous microregions to produce a fully amorphous layer. The index of refraction (1.482 at λ = 575 nm) approaches that of implanted (densified) fused silica (1.475).

Halides

Many of the alkali and alkaline-earth halides are used as transparent insulating materials because of their excellent broad-band transmission, especially in the visible and infrared spectral regions. Of the common materials, the spectral range of transparency of CsBr and CsI is exceeded only by diamond. The response of such materials to irradiation and their defect structures is well-documented and will only be briefly summarized.

The bonding in crystals of these materials is primarily ionic or Coulombic in nature. They generally behave as good electrical insulators with a wide energy gap of 1–12 eV. The point defects carry an effective charge due to the requirement that local electrical neutrality be preserved, thus making them efficient traps for electrons and holes.

The basic intrinsic defects in the alkali halides are cation and anion vacancies and the corresponding interstitial ions. These defects give rise to the familiar color centers. Three kinds of centers have been characterized:

- anion defects
- cation defects
- colloidal defects resulting from agglomeration of the host atoms (or impurity or implanted atoms).

The F-type centers correspond to anion vacancies. The F^+ center contains no trapped electrons and thus carries an effective positive charge. The F and F^- centers contain one or two trapped electrons, respectively. The F_x-centers are clusters of F-type defects. The V-center is a trapped hole on a pair of anions, effectively X^{-2}. Other hole centers include the H-center (a hole trapped along four anions defining a neutral crowdion) and the I-center (hole trapped at an interstitial anion).

These defects can be produced by the nuclear (elastic) damage processes associated with ion implantation. However, the efficiency of the inelastic (electronic) production of defects greatly overwhelms the elastic processes. The efficiency for defect production depends upon the particular alkali halide. At high fluences and at temperatures above the onset of interstitial migration, the stable defect configurations are those containing clusters of point defects, dislocation loops, and often metallic colloids.

Figure 7.3 shows the characteristic optical absorption spectra obtained at room temperature for NaCl, KCl, and KBr implanted with Li^+ (2 MeV, 1×10^{15} ions/cm^2) taken from the data of Davenas.[11] The spectra contain bands due to isolated F- and V-centers as well as aggregates of two and three F-type defects.

Sapphire (α-Al_2O_3)

Sapphire (single-crystalline α-Al_2O_3) is often used for an infrared window material in applications requiring high strength, inertness, or protection against dust or rain drop erosion, or some combination of these. It is transparent from the visible up to almost 7 μm.

Some of the response of sapphire to ion implantation can be directly traced to the bonding and crystal structure. The bonding is primarily ionic but with some covalent character. The oxygen ions are approximately in hexagonal close-packed arrays, and the aluminum ions occupy two-thirds of the octahedral sites. The vacant one-third of the octahedral sites are constitutional vacancies and are necessary to preserve electrical neutrality and to permit the maximum separation of like charges and minimum separation of unlike charges. Any solubility for impurities and any deviation from stoichiometry are so low as to be undetectable. Point defects, thus, act as efficient traps for electrons and holes, producing color centers in an analogous manner as in the alkali halides.

Many features in the optical absorption band have been ascribed to defects associated with the anion (oxygen) sublattice. The F-center (an oxygen vacancy

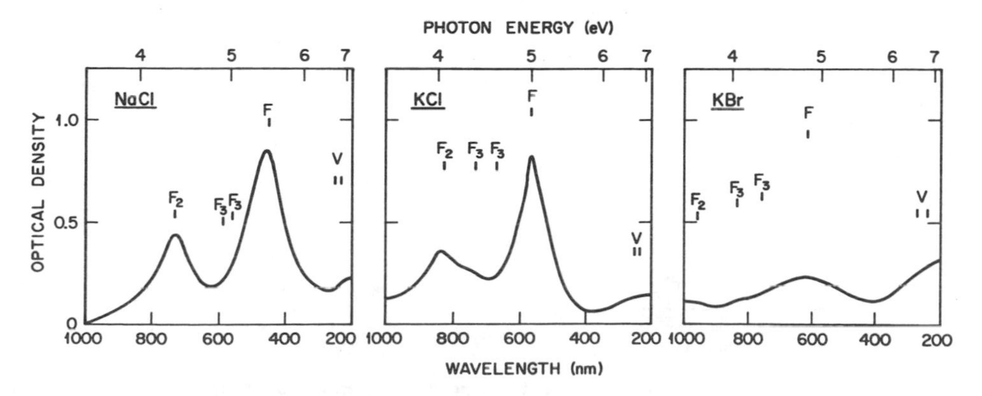

Figure 7.3 **Optical absorption spectra taken at room temperature for NaCl, KCl, and KBr implanted with Li (1×10^{15} ions/cm^2). Absortion bands due to F- and V-centers and aggregates are indicated.[11]**

containing two electrons) and F^+-center (an oxygen vacancy containing one electron) have been identified. Because sapphire has a rhombohedral structure, the F^+ bands are anisotropic and crystal field splitting leads to three distinguishable absorption bands. Aggregates of F-centers have been suggested, but their assignments to features in the absorption patterns are by no means definite.

Centers containing holes (V^- and V^{--}) have been established through electron paramagnetic resonance (EPR) and electron nuclear double resonance (ENDOR) measurements. The hole centers have often been attributed to trapping effects of impurities. A puzzling feature is that their concentrations do not increase during neutron bombardment.

In spite of the large number of studies on the defect structure of sapphire, little is known regarding the nature of the oxygen interstitials or the aluminum interstitials and vacancies.

Ion-implanted sapphire contains large concentrations of dislocations in addition to point defects. Many of the dislocations appear to be stoichiometric interstitial loops and are faulted relative to the cation sublattice.

Amorphous Al_2O_3 is produced by implantation at low temperatures (below ~100 K), where dynamic recovery processes are inhibited. Certain ion species implanted at room temperature also produce an amorphous structure via some "chemical" effect not yet fully understood. There is a significant volume expansion (20–25%) upon amorphization that leads to a change in refractive index of 1.3%.

The effect of ion implantation on the color center formation has been examined for a number of cation species. Figure 7.4 shows the absorption spectra for a sample

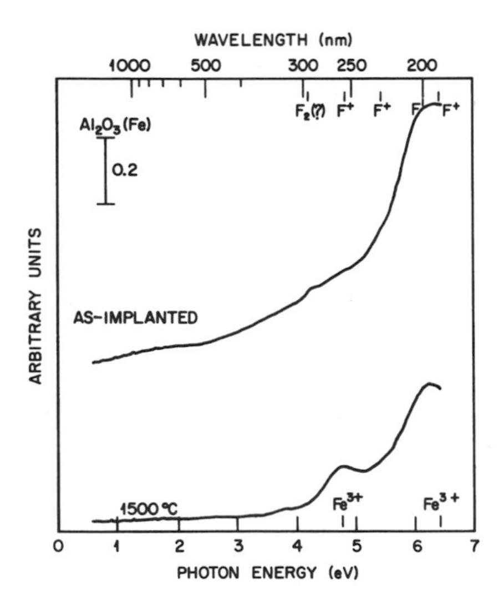

Figure 7.4 **Optical absorption spectra for sapphire implanted with 4×10^{16} Fe/cm^2 and after post-implantation annealing for 1 h at 1500 °C.**

implanted with 4×10^{16} Fe/cm^2 with the ion beam approximately parallel to the *c*-axis. The F-type center assignments are those contained in Table 6.3 of Reference 2. The main features of the absorption curve for the as-implanted sample are consistent with the presence of F-, F^+-, and F_2-centers. At this fluence, the implanted iron is mostly present in the Fe^{2+} state. However, none of the features in the curve could be definitively associated with either isolated Fe^{2+} ions or with iron-point defect complexes. The absorption curve is also shown for this specimen after a post-implantation anneal in oxygen at 1500 °C. All the iron was in the Fe^{3+} state after this treatment. The absorption bands at 4.78 and 6.38 eV have been ascribed to Fe^{3+}.

Figure 7.5 shows the effect of implanting Cr, Fe, or Ga on the transmittance of 1-mm-thick sapphire crystals. There is a small decrease in transmittance in the Cr-implanted specimen (~3% at 1 μm) but significantly larger decreases for Fe at wavelengths shorter than 2.5 μm and for Ga over the entire range. The peak damage energy density was approximately the same for Fe and Ga, and that for Cr was about one-third less. The decrease in transmittance does not appear to scale with deposited energy.

A large number of studies have shown that the near-surface mechanical properties of sapphire are affected by ion implantation. Samples with crystalline implanted surfaces exhibit increases in hardness, fracture toughness, and flexure strength. A large residual compressive stress is also produced. A larger residual stress is produced in samples having the *c*-axis approximately parallel to the ion beam than in those oriented with the *a*-axis parallel to the incident beam.

The increase in flexure strength appears to be primarily due to an effect on the propagation of surface flaws (i.e., due to the compressive stress) and not upon increases in the "intrinsic" strength. Because of the orientation of slip, twin, and

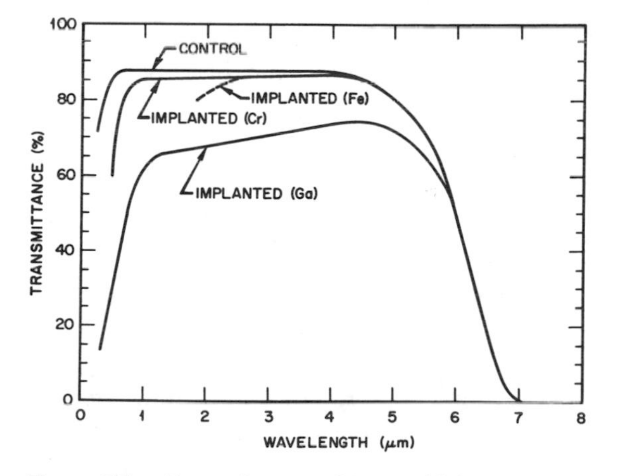

Figure 7.5 Transmittance of 1-mm-thick sapphire crystals (*c*-axis normal to surface) implanted with Cr, Ga, or Fe.

fracture planes relative to the specimen shape, the effect is greater for specimens tested with the tensile axis normal to the *c*-axis than for ones oriented with the *a*-axis normal.

$LiNbO_3$

Several single-crystalline oxide ceramics are useful for electro-optical applications. Single crystals of $LiNbO_3$ are particularly well-suited for fabrication of integrated optical elements and devices. High-optical-quality crystals can be prepared and the material exhibits a strong piezoelectric effect that is accompanied by acousto-optic and electro-optic effects. Both passive devices (waveguides) and active devices (switches, modulators) have been made. Devices are often prepared by the in-diffusion of metals such as Ti, V, or Ni (most commonly Ti). This diffused layer has an enhancement of both the ordinary and extraordinary refractive indices. Devices have also been prepared by using ion implantation: (1) to produce a damaged region having a refractive index different from the matrix or (2) to dope with Ti to a relatively high concentration (>10 at %) in thin layers.

The structure of $LiNbO_3$ is related to the perovskite oxide structure. The large oxygen ions form close-packed layers and the smaller Li^+ and Nb^{5+} ions are regularly distributed into the octahedral sites. One-third of these octahedral sites remain empty, given the structural dipolar symmetry along the *c*-axis: Li–Nb–vacancy–Li–Nb–vacancy–. . . .

The major intrinsic defect is considered to be the oxygen vacancy. The concentration of this defect is a strong function of impurity content and reduction state (i.e., oxygen partial pressure). Cation vacancies may also be present.

Optical absorption bands induced by low-temperature irradiation or by thermochemical reduction treatments have sometimes been ascribed to F- or F^+-centers. These assignments are not unambiguous, however.

The first studies of ion-implanted $LiNbO_3$ relied on the damage produced by light-ion bombardment to cause an expansion in the lattice and thus a change in the refractive index. A strip waveguide could be produced by concentrating the damage (reduced index) in a deep layer surrounded on both sides by undamaged matrix.

The rate of production of damage (defects), as determined from RBS measurements, is strongly dependent upon the orientation of the ion beam relative to the crystallographic axes. Götz[3] has divided the amounts and type of damage, changes in volume, and changes in optical properties into three ion fluence (or damage energy) regimes. For low-fluence N-implantation ($<1.5 \times 10^{15}$ ions/cm^2), point defects are the main manifestation of damage. These appear to be rearrangements of Nb ions onto vacant octahedral sites or onto Li sites. There is a strong increase in damage with increasing fluence up to about 5×10^{15} ions/cm^2. The clustering of point defects leads to the completely random distribution of Nb relative to the *x*- and *y*-axes. Some degree of order persists in the *z*-direction. The niobium ions are completely randomly distributed for higher fluences.

For crystals with *x*- or *y*-cut, the change in volume for deposited damage energies below about 10^{21} keV/cm^3 is below the limit of detection for the technique used in these studies (~1%). The volumc increases with damage energy (or fluence) until a saturation value of 11% is reached at ~5×10^{22} keV/cm$_3$.

Both the ordinary and extraordinary refractive indices decrease with increasing ion fluence. A maximum decrease of ~7% (λ = 632.8 nm) is observed at 4×10^{16} N/cm^2. Decreases in the refractive indices are detected at fluences below those necessary to cause detectable volume changes. In this regime, the change in index should be given by the variation in polarization. At the higher fluences, the changes are due to the change in density of electrons due to the volume dilatation, change in polarization, and the effect of changes in crystal structure on local fields.

The damage introduced in the low-fluence regime can be completely removed by annealing 30 min at 570 K. An annealing temperature of 1370 K is required to remove the high levels of damage characterized by the random distribution of Nb. All heavily damaged crystals exhibit anomalous behavior with respect to the RBS measurements upon annealing. The amount of disorder measured by RBS appears to increase during the annealing step at 570 K. This change in backscattered yield is due to a change in composition (and density) of the near-surfacc rcgion caused by a loss of Li. This loss of Li has little effect on the ordinary index of refraction but a large effect on the extraordinary component.

Recently it has been shown that direct ion implantation of Ti followed by thermal annealing introduces high concentrations of Ti into substitutional sites in the $LiNbO_3$ lattice.[5] This procedure has been used to fabricate both planar and channel waveguides with low optical attenuation. Direct implantation produces higher Ti concentrations in a more shallow surface layer than does thermal diffusion of Ti.

Initial studies indicated that ion implantation often leads to surface decomposition of the $LiNbO_3$ duc to the out-diffusion of Li. This surface decomposition can be avoided by implanting at 77 K followed by immediate thermal processing or by storagc at 77 K.

A post-implantation thermal treatment is necessary to remove the ion beam-induced damage and to cause full incorporation of the Ti onto lattice sites. The implanted region is amorphous after the high fluences required to give the desired Ti concentration (2.5×10^{17} Ti/cm^2 at 360 keV to give ~10 at % Ti). Annealing 1 h at 1275 K in oxygen saturated with water vapor almost completely removes the lattice damage without producing Ti precipitates. The recovery of the damaged region occurs by solid-phase epitaxy (SPE) from the undamaged, underlying matrix.

Preparation of Optical Components by Ion Implantation

Preparation of optical waveguides by ion implantation has been demonstrated for amorphous silica, α-quartz, sapphire, $LiNbO_3$, etc. For details of preparation, see Götz[3] and Townsend.[4] In amorphous silica, the optical beam is confined by total internal reflection in a region of high refractive index generated by densification of the SiO_2-tetrahedral network. Chemical changes by high-fluence nitrogen implantation that forms oxynitrides can give additional index enhancement.

Optical waveguides in crystalline oxides (α-quartz, sapphire, $LiNbO_3$) rely on an "amorphous" or highly disordered region of lower density and refractive index than the adjacent matrix region. The result is a high-index region (crystalline waveguide) surrounded by a low-index amorphous optical barrier. Color centers anneal out before regrowth of the amorphous material begins, thus, the damaged crystalline regions can be repaired by a low-temperature anneal.

Materials of interest for integrated optics (α-quartz, $LiNbO_3$, $LiTaO_3$) must be used in the crystalline state to make electro-optic modulators or surface acoustic wave devices. Tight control of implantation parameters for light-ion implantation can give sufficient lattice dilatation to give waveguiding without amorphizing the region. Bremer[4] has been able to produce a multimode waveguide by implanting 2 MeV He^+ into $KNbO_3$. The damaged but crystalline surface did not require thermal annealing.

In the case of Ti-implanted $LiNbO_3$ the amorphous implanted region can be removed without affecting the Ti distribution by solid-phase epitaxy regrowth from the substrate. This procedure has been used to fabricate single-mode channel and planar waveguide structures as well as Mach–Zehnder type waveguide modulator.[5]

The effect of ion implantation on the mechanical properties of sapphire has been used in the production of large infrared windows for certain applications that expose the devices to very large stresses and to rain and dust erosion. The increase in flexure strength due to implantation gives higher reliability at higher design stresses than is possible with the best optical polishing techniques. It should be possible to increase the strength of other potential window materials in a similar manner.

Townsend[4] has suggested that it may be possible to produce solid-state lasers in a large number of inorganic materials by using ion implantation for doping or to alter the carrier density in various regions.

It is apparent that ion beam technology offers much promise for fabricating optical devices. The process is characterized by its homogeneity and reproducibility and has the possibility of generating fine-patterned structures. The potential is limited only by the imagination of the designers of devices.

References

1 J. S. Williams and J. M. Poate. *Ion Implantation and Beam Processing.* Academic Press, New York, 1984.

2 F. Agullo-Lopez, C. R. A. Catlow, and P. D. Townsend. *Point Defects in Materials.* Academic Press, New York, 1988.

3 *Ion Beam Modification of Insulators.* (P. Mazzoldi and G. Arnold, Eds.) Elsevier, Amsterdam, 1987.

4 *Structure–Property Relationships in Surface-Modified Ceramics.* (C. J. McHargue, R. Kossowsky, and W. O. Hofer, Eds.) Kluwer Academic Publishers, Dordrecht, the Netherlands, 1989.

5 C. W. White, C. J. McHargue, P. S. Sklad, L. A. Boatner, and G. C. Farlow. *Materials Science Reports.* **4**, 41, 1989.

6 P. D. Townsend and F. Agullo-Lopez. *J. Phys. (Paris) Colloq. C6.* **41**, 279, 1980.

7 G. W. Arnold. *IEEE Trans. Nucl. Sci.* **NS-20**, 220, 1973.

8 E. P. EerNisse. *J. Appl. Phys.* **45**, 167, 1974.

9 R. L. Hines and R. Arndt. *Phys. Rev.* **119**, 623, 1960.

10 A. P. Webb and P. D. Townsend. *J. Phys. D.* **9**, 1343, 1976.

11 J. Davenas. Ph.D. dissertation. University of Lyon, 1972.

8

Laser-Induced Damage to Optical Materials

MICHAEL F. BECKER

Contents

8.1 Introduction

The objective of this chapter is to identify and explore some important applications of materials characterization to the study of laser-induced damage to optical materials. In keeping with the theme of this volume, we restrict consideration to laser damage of surfaces and do not consider the characterization of bulk damage. Furthermore, the examples cited are all of pulsed laser-induced damage studies. In some cases, the described techniques may also apply to continuous-wave (CW) laser damage studies, but such damage is not as commonly reported as a significant constraint to optical system design.

We begin this chapter with basic definitions of laser damage and the statistical considerations in measuring laser-damage thresholds. At this point, techniques for diagnosing predamage sites and changing the damage threshold are reviewed. Actual damage diagnostics, considered next, are grouped into two categories: in situ diagnostics (collected during the damaging event) and postmortem diagnostics collected from the sample after laser damage has occurred. The chapter concludes with a brief discussion of future directions in the characterization of laser-induced damage.

8.2 Laser Damage Definition and Statistics

The purpose of this section is to give some background on laser damage, how it is defined and measured, and indications of what surface diagnostics might be useful in its study. Immediately one encounters two unique difficulties in the study of laser damage. Over the years, the definition of laser damage has caused difficulty for the scientists and engineers studying this phenomenon. It is unfortunate that the definitions tend to involve subjective criteria and require statistical analysis to give a meaningful measure of damage threshold. Applications-driven definitions, such as "a mirror is damaged when its reflectance drops by x% when used in its intended optical system," although easy to quantify, are not useful in the scientific study of the causes and prevention of laser damage. By the time a mirror is deemed damaged by a functional definition, the microscopic evidence of the damage onset has been obliterated. This relates directly to the second fundamental problem in laser damage study. The damage process, by its very nature, is a phase transition that is sometimes catastrophic and erases evidence of its initiation, which one would have liked to study by employing materials characterization techniques. We start our discussion of laser damage by addressing the first of these difficulties.

Defining Damage

The most common and widely applicable (although perhaps not always the best) definition of laser damage is the detection of any change in the irradiated surface as observed before and after measurement, usually with optical Nomarski or dark-field microscopy. Although one might expect laser damage to occur at the front surface of an optical element, this is not always the case. For transparent materials functioning as windows, laser damage is often initiated at the exit surface of the sample. The reason, first pointed out by Crisp et al.,[1] is that, due to the Fresnel reflection coefficients, the electric fields of the forward and backward traveling waves add in phase at the transition to a lower index material (back face) and add out-of-phase at the transition to a higher index material (front face). Damage normally initiates at the back face if the majority of the laser power is able to reach the back face of the sample and if there are no easily damaged thin films on the front surface.

This definition introduces an observer-dependent bias which has been addressed by either statistical analysis techniques or by technological solutions for more narrowly defined situations. In the latter case, two techniques seem to have merit. The first is to correlate damage on thin-film optical coatings with an increase in scattered light from the irradiated spot as detected by a CW probe laser. Stewart and Guenther[2] found near-angle optical scatter to be the best indicator of damage on thin-film samples. Near-angle scattered light could be detected at the 10–100-ppm level and allowed damage detection to be automated. A second damage detection technique was developed by Penzoldt et al.[3] for application to laser surface damage studies on CaF_2, LiF, and MgF_2 samples. In their method, the shock wave propagating in the gas above the sample surface is detected by a CW laser deflection technique. They found that a sudden increase in the shock-wave signal coincided with or began at a slightly lower fluence than did a reduction in specular reflectivity at that site. Not only did this technique work for these fluoride samples, but they also applied it to polymethyl methacrylate (PMMA), Al, and Cu surfaces.[4]

Collecting Damage Statistical Data

The current technique used for collecting statistical laser-damage data, first suggested by Foltyn,[5] is to irradiate a number of sites on a sample with one (or *N*) laser pulses at fluences ranging from below that where no sites damage to above that where all sites damage. Fluence is taken to be the peak value of energy per unit area within the beam spot, or the peak-on-axis value for a Gaussian laser beam. For one pulse per site the data are termed 1-on-1, and for *N* pulses per site the data are termed *N*-on-1. The latter case is discussed further in the section "Changing the Damage Threshold," where changes in the damage threshold induced by multiple pulses are considered. Following irradiation, the sites are examined and the adopted definition of damage is applied to each site, then assigned to either the damaged or undamaged category. Then, the fluence range is divided into increments containing generally five or more sites, and the probability of damage for a site irradiated in that range is set equal to the fraction of sites damaged. These data comprise the measured probability curve, plotted as damage probability versus peak laser fluence as shown in Figure 8.1 for typical thin-film high-reflector mirrors. The onset or

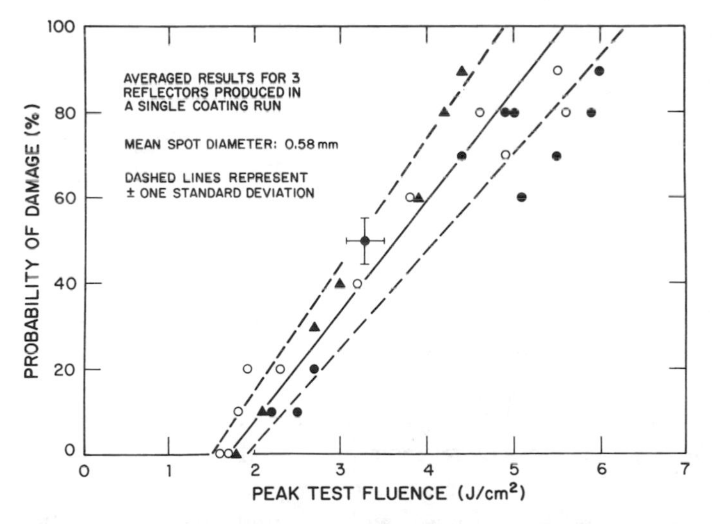

Figure 8.1 **Results of testing three dielectric thin-film reflectors with a spot diameter of 0.58 mm. The dashed lines extend ±1 standard deviation from the best fit. The error bars indicate a ±5% uncertainty in measurement of both fluence and probability. (From Foltyn.[5])**

threshold fluence is extracted from the measured probability curve via fitting the data to an assumed form; generally, a simple linear regression is used. The zero-probability intercept for the fitted curve is then defined as the damage onset. Complications arise when a model-based form for the probability curve is assumed or when the samples might have defects with a distinctly bimodal distribution of damage onsets. These aspects are treated next.

Types of Damage Probability Distributions

The technique already described for plotting damage probability versus fluence was developed by Foltyn[5] in order to explain the variations in damage threshold and slope of the probability curves for damage measured for different Gaussian laser spot sizes. The resulting theory postulated a density of defects with a constant damage threshold lower than that of the surrounding film material. Large fluctuations in the damage measurements and decreasing slope of the damage probability curves were associated with a defect density sufficiently low that the probability of finding a defect near the center of the laser beam spot was significantly less than one. By determining the extrapolated zero probability intercept, or onset, researchers could eliminate the variation due to beam size. Further development of this theory of defect-dominated laser damage has included the addition of a variety of types of defect ensembles[6, 7] and closed-form solutions for defects at both surfaces and in the bulk.[8] Data reported by these authors on thin film coatings and bulk polymers have been in agreement with this defect-ensemble model.

Identification of Pre-Damage Sites

In the case of surfaces whose laser damage is dominated by defects as described previously, the salient experimental question becomes "Can those defect sites which will lead to damage be identified in a nondestructive manner?" Low-power optical measurements of scatter and absorption give no indication of which observed scattering or absorbing centers will eventually be damaged. More successful attempts at identifying predamage sites involved the use of subthreshold pulsed laser illumination. Initial experiments involving video imaging of the optical scatter of the subthreshold pulses proved to be inconclusive.[9, 10] More sophisticated experiments which added measurement of the visible emission spectrum of individual defects were equally unable to identify a characteristic signature for sites prone to damage.[11]

The most successful technique to date for predamage defect identification is designed to observe directly the transient temperature rise of defect sites in the irradiation area of subdamage threshold pulses. This technique, developed by Clark and Emmony,[12] utilizes a dye laser synchronized with the subdamage laser to illuminate stroboscopically the target site in a Schlieren imaging system. This rather complex system is diagrammed in Figure 8.2. Images before irradiation, of transient heating, after cooling, and after damage were acquired for comparison. Good

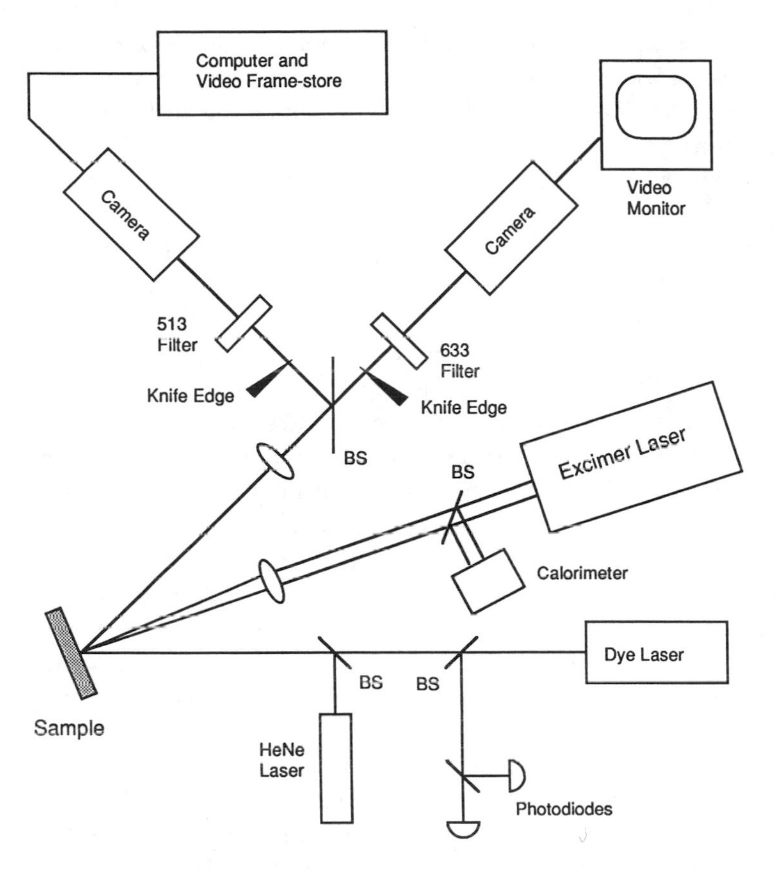

Figure 8.2 Experimental setup for viewing predamage sites. (From Clark and Emmony.[12])

correlation was obtained between those sites showing significant heating and those eventually damaging for a wide range of materials: Al mirrors, Ge surfaces, and dielectric thin-film reflectors. The process of identifying predamage sites nondestructively is still not optimized and is the subject of current research efforts.

Changing the Damage Threshold

The damage threshold has been found to be a function of the history of an optical element. In fact, the optical irradiation history plays an important role in determining the damage threshold at later times. Depending on the material, it is possible to raise the damage threshold through the processes of conditioning with prepulses or polishing with CW laser irradiation or to lower the threshold through the process of accumulation of damage due to the previous pulses. Generally, conditioning with prepulses or polishing are applied to dielectric thin films and bare dielectric material surfaces while accumulation is observed on bare metal and semiconductor surfaces (and in bulk damage of dielectric and polymer materials). These experiments have drawn from a wide range of surface diagnostics, but understanding has come slowly because of the wide variability in the perfection of the optical surfaces and coatings to be treated.

Laser conditioning Conditioning can take on a multitude of forms since an infinite number of different series of prepulses can be applied to a test site before it is damage-tested. Fortunately, only a few types of conditioning schemes are in common use, and of these, a linearly ramped sequence of pulses has been as good or slightly better than other techniques.[13] Unfortunately, it is difficult to treat large surface areas with this technique, so other less effective methods, such as beam rastering,[14] have been applied to condition large areas. Due to the wide variation in sample quality and preparation, conditioning does not always occur, or in other cases, it may show extreme site-to-site fluctuations. Conditioning studies may generally be divided into several subareas: irreversible conditioning, reversible conditioning or cleaning, and CW laser conditioning or polishing.

Conditioning has been observed for cases where larger defects, visible by optical scattering, were suspected as the damage cause and for cases where invisible defects were postulated as the cause of damage. In the former case, Kerr and Emmony[15] used excimer laser pulses at a wavelength of 248 nm to condition bare surfaces of SiO_2, CaF_2, GaAs, and Al. The conditioning effect increased damage thresholds markedly for the first two and somewhat for the latter two materials, and resulted in the visible removal of surface defects and scratches in the treated region. The possible relief of stress was also mentioned as a possible cause for conditioning. These results are very similar to those to be described in a following section for CW laser polishing of surfaces. These experiments are also the first observation of an increase in damage threshold with multiple pulses for metal or semiconductor surfaces. They usually undergo a decrease in threshold due to accumulation in multipulse tests.

In the case of permanent conditioning of thin films without visible defects as the cause of damage, the laser-damage group at Lawrence Livermore National Laboratory (LLL) used an array of analysis techniques to eliminate several possible mechanisms for conditioning and suggest yet another possibility. In a series of papers on laser damage to various types of thin-film coatings used in fusion laser systems (at 1.06 μm wavelength and its second and third harmonics),[13, 14, 16, 17] they reported permanent changes in threshold not related to film water content. Furthermore, the effect was found to be independent of the number of film layers and any measurable recrystallization within the films. Electron paramagnetic resonance experiments on SiO_2 films gave no indication that filling of midgap states and thereby removing easily ionized electrons were involved. There was a positive correlation between the emission of energetic neutral particles from the film surface and conditioning. A typical neutral flux time-of-flight waveform is shown in Figure 8.3 for Hf neutrals emitted from an HfO_2 film surface. The authors propose that the

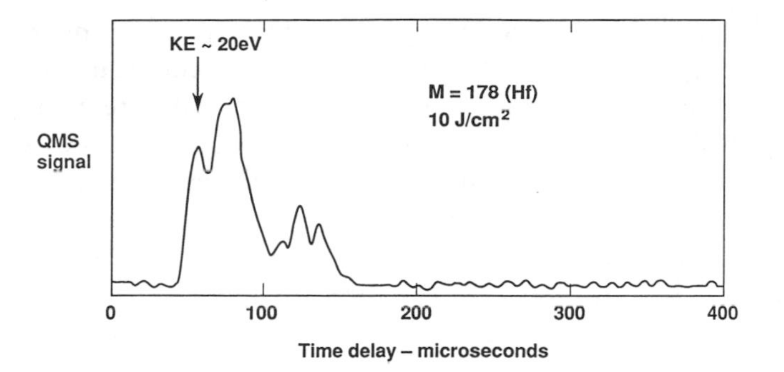

Figure 8.3 **Typical time-of-flight signal for Hf neutrals from an HfO_2 thin film at the emission threshold for this site. (From Schildbach et al.[17])**

energetic emission of neutrals is associated with fractoemission from microcracks resulting from the stress relief at defect regions which suffered more heating than the surrounding material.

Reversible conditioning has been associated with the water content of films and surfaces. Although it has been definitely ruled out in the preceding experiments, Arenberg has reported damage experiments on thin films that definitely implicate water in the temporary conditioning he observed.[18] Further experiments showing conditioning at 1.06 μm and no conditioning at 0.53 μm further indicated the involvement of water.[19]

Laser polishing of surfaces and thin films, primarily with high-power CW CO_2 lasers, has shown generally mixed results, although damage threshold improvements that were observed could generally be correlated with the removal or minimization of surface defects and scratches. Initial experiments on laser-polished fused silica surfaces by raster scanning indicated that laser polishing improved the damage resistance of thin-film coatings applied to them.[20, 21] Later experiments at the Air Force Weapons Laboratory (AFWL) on both bare fused silica surfaces and thin films utilized flood illumination of the entire optic for conditioning. Although reduced surface roughness and optical scatter, as well as increased stress, were always observed, increased laser-damage resistance was sporadic.[22] No change in surface absorption at 351 nm was observed in similar polishing experiments.[23] When thin films were laser-polished and then damage-tested, many optical and crystallographic changes were observed but only several film types showed limited conditioning, and the conditioning was not correlated directly with more consistent changes in the film microstructure and related parameters.[24]

Accumulation Accumulation is the reduction in the damage threshold for multiple pulses applied to a single site at the same fluence. In the context of pulsed laser-induced damage, accumulation usually implies a pulse repetition rate low enough that the damage site is not significantly heated by the average laser power;

in other words, the site has cooled back to ambient by the time the next laser pulse arrives. The challenge to researchers has been to determine what is accumulating in the material to account for the damage threshold reduction. It is necessary to point out that if damage is a probabalistic process, an apparent accumulation due only to the multiplication of probabilities can result in a reduction in the experimentally measured damage threshold as pointed out by Fry et al.[25] For most experimental data on materials showing accumulation, the reduction in damage threshold is more than could be accounted for by probabalistic arguments alone.

For metal surfaces, the cause of accumulation has been well explored. Initially, Musal[26] proposed to apply thermomechanical techniques to the laser damage of metals and determined the onset of yield or plastic deformation. Subsequent experiments confirmed the involvement of plastic deformation and slip in the damage of metals[27, 28] and established an empirical equation describing the reduction in damage threshold after N pulses[29]: $FN = F_1 N^r$, where r is the slope of the log–log plot of damage fluence versus N and F_1 is the single-pulse damage onset fluence. Typical data for N-on-1 damage to metal surfaces showing accumulation are presented in Figure 8.4. In the figure, the Cu data follow the accumulation equation exactly, whereas the Al curve deviates for low pulse numbers because local defects are dominating damage at this fluence rather than bulk thermomechanical fatigue. Accumulation in metals is on firm theoretical ground now that the accumulation behavior has been related to the fatigue behavior of bulk metals[30] using the photothermal displacement technique described in the next section.

The situation is not nearly so well-understood for accumulation at semiconductor surfaces. Accumulation is commonly observed for damage with picosecond pulses and usually observed for nanosecond pulses for a variety of materials: Si, Ge, and GaAs.[31, 32] Slip or plastic deformation has never been observed in the laser damage of semiconductors, and experiments at elevated temperatures and with a wide range in pulse repetition rates have indicated that the accumulation is permanent.[33]

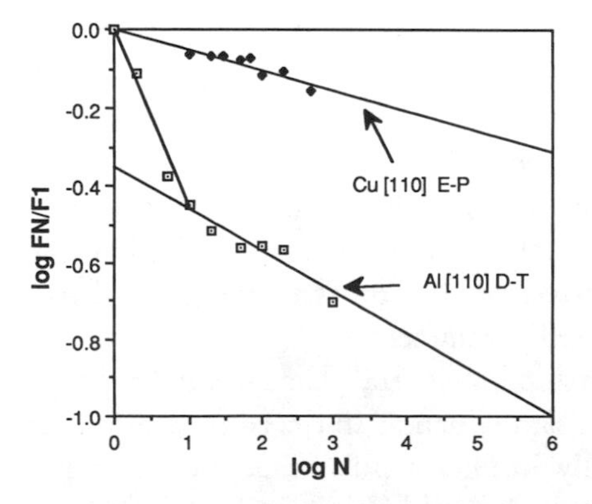

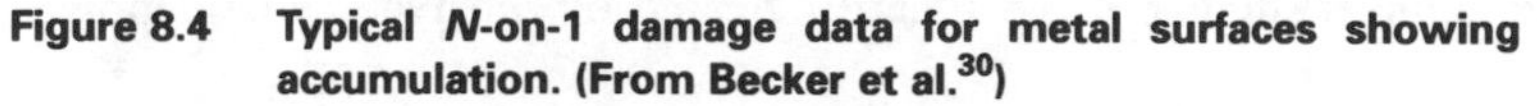
Figure 8.4 **Typical *N*-on-1 damage data for metal surfaces showing accumulation. (From Becker et al.[30])**

As yet, no specie of defect has been identified to account for the observed accumulation in semiconductors.

8.3 In Situ Diagnostics

Probably the best way to learn about a transient, nonequilibrium process is to observe during the process and not just analyze the initial and final states. The objective of in situ diagnostics in laser damage is to collect information on the dynamic damage event as it occurs. Of particular interest in the study of laser-induced damage are measurements of the thermal and thermomechanical behavior of the sample and the energetics and species of particles ejected from the surface during damage. We start with a description of the former family of techniques.

Photothermal Techniques

Photothermal techniques are a group of techniques based on exciting the target surface with a pulsed (or chopped CW) laser and observing either a thermally induced change in the surface itself (reflectance, curvature, or slope) or the secondary change in the heated air above the sample surface (thermal lens, shock wave, or mirage effect). Although the case of chopped CW excitation is not directly useful for in situ studies of damage, it is widely used to measure the thermal transport properties of optical thin films and surfaces.[34–37] With pulsed excitation, transient thermal transport properties can also be measured but with a smaller signal-to-noise ratio. Thus, the technique is not as desirable for material properties measurement as it is for resolving transient changes at the surface.

The first suggestion to use a photothermal technique as a noninvasive method of probing the peak temperature reached by a laser irradiated surface was made by Bailey et al.[38] Their technique used detection of the thermal lens effect and a relatively low-power pulsed argon ion laser. This represented a significant advance over thermal emission techniques for temperature measurement which are generally slower in their response but very useful in the study of CW laser-induced damage.[39] The photothermal displacement technique (PDT) used today was introduced by Olmstead et al.[40] for spectroscopic measurements of doped glass. The Gaussian profile excitation laser beam induces a similarly shaped thermal expansion of the sample surface. The transient expansion is probed by a He–Ne laser beam focused at the region of maximum surface slope, as shown in Figure 8.5. The deflection of the probe beam is proportional to the slope and height of the heated region. Given a sufficiently fast detection system, the thermomechanical history of the pulsed excitation event can be accurately recorded. A typical experimental setup for performing these measurements is shown in Figure 8.6 from a paper in which Karner et al. measured thermal transport in metals.[41] The theoretical foundation of PDT is completed by the analysis of PDT and competing mechanisms by Opsal et al.[42] Although their experimental application involved modulated CW laser excitation, their results can be applied to the case of short-pulse excitation as well.

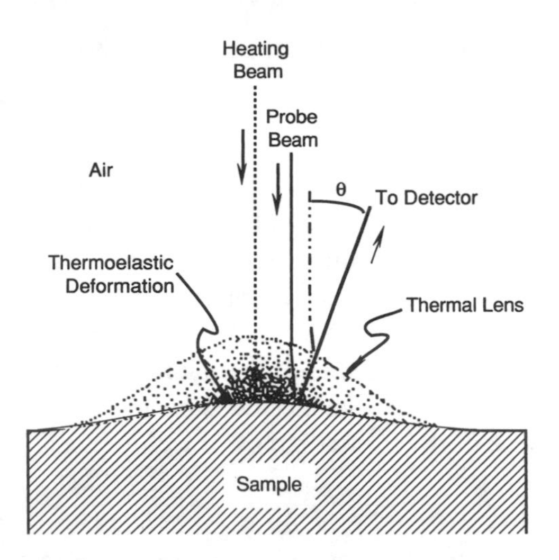

Figure 8.5 Diagram of probe beam deflection at a laser-heated surface site. (From Opsal et al.[42])

A wide variety of applications have been found for PDT in the study of surfaces, thin films, and laser damage. The technique is even useful on transparent materials[43] and polymers with UV laser excitation.[44] An excellent overview of these applications can be found in volume 47 of *Topics in Current Physics.*[45] In this section, we will concentrate on example applications of PDT to laser damage of metal mirrors and thin-film coatings.

Metal surfaces A good example of the use of PDT for nondestructive measurement of thermomechanical surface displacement and indirectly of surface temperature is the study of laser damage to Mo metal surfaces conducted in the author's laboratory.[30, 46] Because metals are nontransmissive, the thermal source term is at the surface, and the surface will have the highest temperature. Combined with an estimate of the heating depth or, better still, a transient-heating computer model for the material, surface displacement measurements can be converted into surface temperature values. In our work on Mo, the objective was to characterize the peak surface displacement and temperature rise during a single pulse event, before further study of accumulation in multipulse damage. The experimental setup was similar to that in Figure 8.6 where a 10-ns pulsed Nd:YAG laser at 1064 nm was the excitation source, and the knife edge and photodetector were replaced by a fast bicell detector and differential amplifier.

Typical results for the peak surface displacement of Mo are shown in Figure 8.7. The data presented were derived from the raw surface-angle data by use of the Gaussian excitation-beam geometry. Signal-to-noise considerations limit the sensitivity of the instrument to about 1 nm of surface displacement. This sensitivity is only possible if high-pass filtering and vibration isolation are utilized. The data for Mo usually showed a marked change for fluences above the surface damage onset;

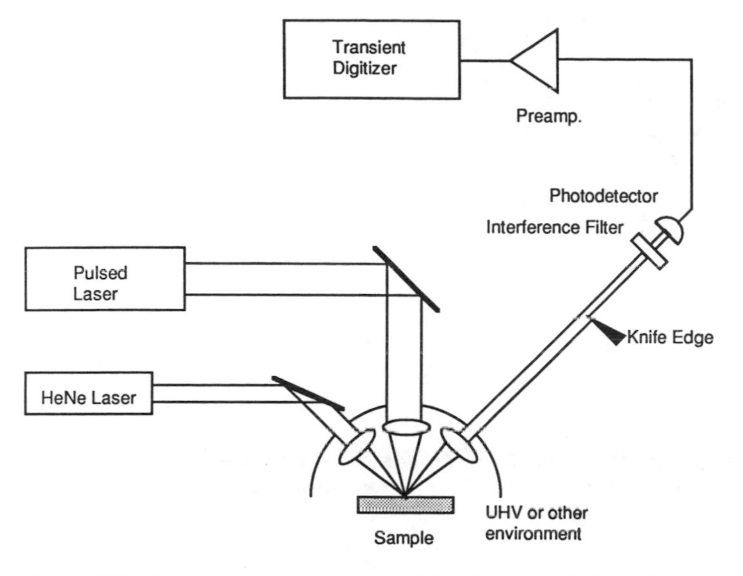

Figure 8.6 Experimental arrangement for the pulsed photothermal displacement technique. (From Karner et al.[41])

both the scatter and slope would increase. Figure 8.8 shows the same data converted to peak surface temperature using a simple estimate of heating depth and a constant value of thermal expansion coefficient. For comparison, computer model results are also shown using known temperature dependent material parameters for Mo with no fitted parameters. In this case, the agreement is quite good. In general, computer modeling can be used to check the estimate of the heating depth and influence of the temperature dependence of the thermal expansion coefficient. In the case of Mo and most other metals such as Cu and Al,[47] the temperature rise was found to be

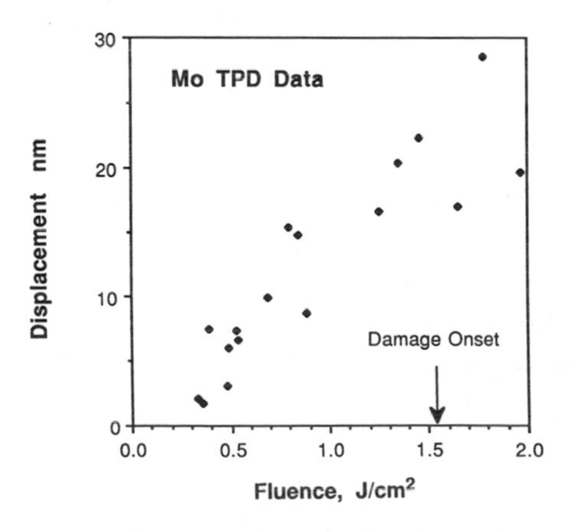

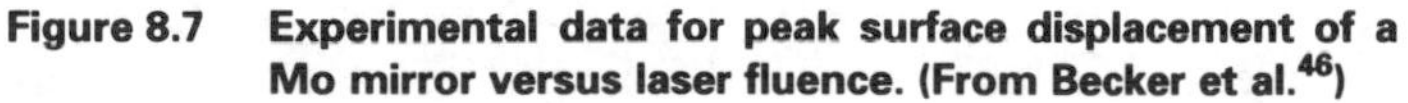

Figure 8.7 Experimental data for peak surface displacement of a Mo mirror versus laser fluence. (From Becker et al.[46])

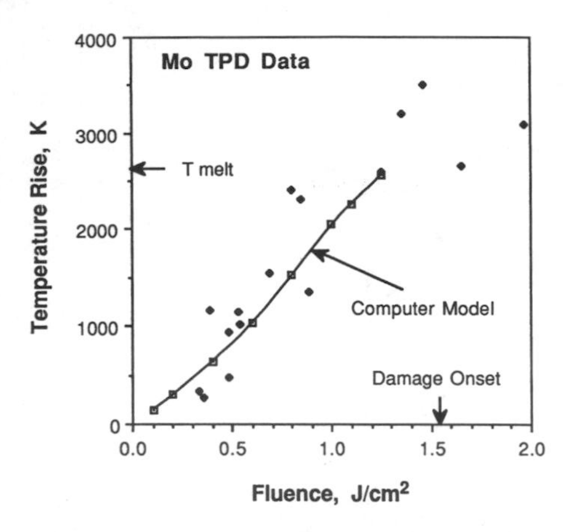

Figure 8.8 Comparison of numerical model (*solid line*) with experimental data (*filled diamonds*) for peak surface temperature rise versus laser fluence for a Mo mirror. (From Becker et al.[46])

roughly linear with laser fluence, and surface displacements were measurable considerably below the onset of surface damage.

Thin films Although thin films in many cases are not strongly absorbing, and the peak temperature during pulsed-laser irradiation does not necessarily occur at the surface or even in the top film layer, PDT is a useful technique in the study of thin-film damage. Wu et al.[48] have measured thermal expansion coefficients of dielectric thin films using the PDT technique. They found it useful for characterizing thin-film properties which may be different than those of the bulk material.

Perhaps the most intriguing application of PDT to thin films is its ability to locate the depth of absorbing defects and damage locations within a stack of thin-film layers.[49, 50] By recording the surface deformation transient and applying time-delay analysis, researchers can determine the locations of multiple thermal sources within the stack. This is shown schematically in Figure 8.9. Damage is clearly indicated by an orders-of-magnitude increase in the PDT signal. At fluences below the damage onset, the PDT signal is dominated by thermal-wave effects rather than by shock or photoacoustic waves. Film layers with large thermal increases can easily be identified. Wu and co-workers also demonstrated a second important application of PDT by determining the evolution of the absorption location during a series of laser pulses used to condition the test site. Not only the damage location, but also changes in subthreshold heating, could be followed on a shot-to-shot basis. The full potential of this technique for the investigation of laser damage of thin films has yet to be explored.

Particle Emission

The emission of charged and neutral particles has long been viewed as a means to study the interaction energetics of a laser beam with a material surface. Early work

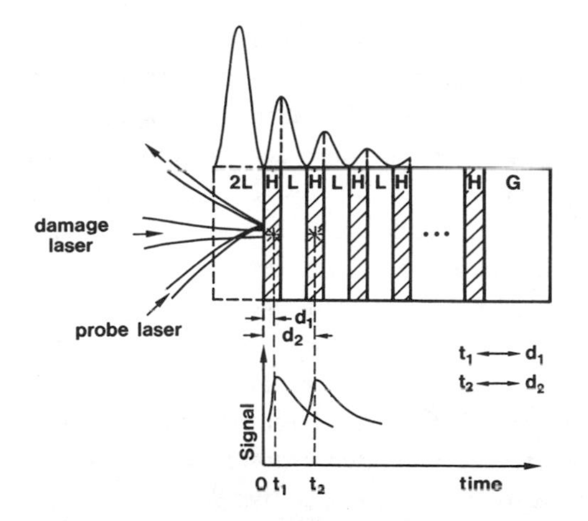

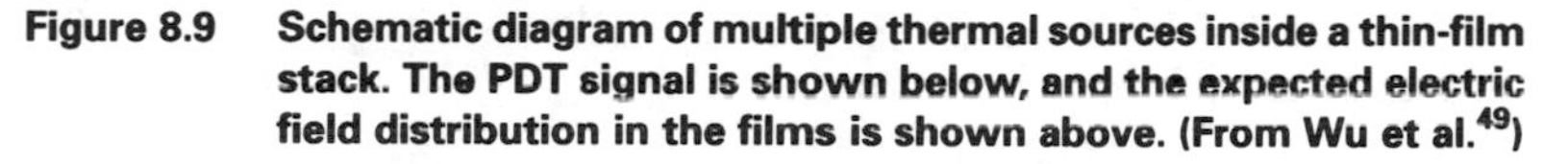
Figure 8.9 Schematic diagram of multiple thermal sources inside a thin-film stack. The PDT signal is shown below, and the expected electric field distribution in the films is shown above. (From Wu et al.[49])

focused on detecting the strong charged-particle emission associated with surface damage.[51, 52] It was soon realized that there was an even stronger flux of neutrals associated with the irradiation event and that with more sensitive detection, both charged particles[53] and neutrals[54] could be detected below the onset of damage. An important distinction must be made here between the emission of surface constituent particles and surface cleaning due to the removal of contaminants such as water and carbon or hydrocarbons. As mentioned previously, surface cleaning has in some cases been associated with conditioning of thin films and surfaces. There is wide agreement as to how surface cleaning is accomplished and that it can be done without damage using either pulsed or CW irradiation.[55]

The clearest link between particle emission and laser damage has been established for the subthreshold emission of constituent particles. A number of interesting analytical techniques have been applied to determine the mass and energy distributions of the emitted neutrals.[54, 56] Although many of these techniques derive from the study of atomic emission from cleaved material surfaces in ultra-high vacuum (UHV), the physics of the emission process is considerably different for real surfaces and thin films, even if they are thermally or laser cleaned in UHV. The work of the LLL group has been most enlightening in this respect.[17, 54] They conclude that there is undoubtedly a correlation between neutral emission onsets and laser damage onsets indicated by observations on a wide range of material and thin-film surfaces. In clean, well-prepared samples, there was emission only of constituent species. This was true even for thin films, in which visible defects that often led to damage were found to contain no impurities when examined by laser ionization mass analysis (LIMA).

The energetics of emission have been analyzed by time-of-flight analysis. Certain common characteristics, shown in Figure 8.3, were observed frequently: a high-energy peak at tens of electron-volts energy and Maxwellian or more complex peaks corresponding to much lower energy emission. Temperatures assigned to the Maxwellian peaks ranged from 300 to 800 K. These observations answered one question, but raised another. For the case of NaF, the high-energy emission was associated with the appearance of visible cracks and fissures. In other materials showing this peak, including HfO_2 thin films, fractoemission from microcracks was a reasonable explanation for the high-energy neutral emission. The problem arises in the interpretation of the cooler emissions whose Maxwellian temperature may range from only tens to hundreds of degrees above ambient for irradiation near the emission detection threshold. In ZnS, this low-temperature emission is not accompanied by a high-energy peak. A thermal mechanism has been ruled out, and the experimenters have postulated an electronic mediated atomic emission event akin to surface ablation but initiated by below band-gap photons. Whether any of these emission effects are localized on the surface is unknown at this time, but emission imaging experiments have been proposed.

8.4 Postmortem Diagnostics

The usefulness of postmortem damage diagnostics lies in the knowledge to be gained about the new state or phase of the material after the damage event. As pointed out previously, knowledge of the details of a catastrophic damage event may not always be discernable from postmortem analyses. However, postmortem diagnostics are indispensable in the study of material changes taking place during conditioning and multipulse laser damage experiments. Two examples have been chosen for examination in this chapter: surface charge on the optical surface and surface structure analysis. The first relates directly to the preceding diagnostics on charged and neutral particle emission as the irradiation site is the source of charge deposited on the sample surface. The latter example relates more directly to the study of microscopic and catastrophic structure changes which occur during pulsed laser irradiation.

Surface Charge State

It is evident from the experiments described in the section "Particle Emission" that neutral and charged particles are emitted in quantity during a laser-damage event and often in smaller, but not insignificant, quantities for irradiation below the damage onset. My coworkers at AFWL and I developed a technique whereby the charge state of a dielectric surface might be mapped with 0.5-mm resolution. The apparatus[57] was designed to measure not only surface charge, but changes in surface potential in metallic and semiconducting samples, as well. The most significant results were obtained for dielectric surfaces and thin films. The principle of the measurement was that of a vibrating-plate capacitor. A probe tip, 0.5 mm in diameter, was vibrated at a mean height of 70 μm above the sample surface. The tip potential was adjusted using feedback to null the ac component of voltage induced on the tip. The null dc potential of the tip was equal to the surface potential (including free charge if it was present). Sensitivity for this device was below 0.1 pC/cm^2. The tip was scanned across the sample surface before and after laser irradiation to measure the change in surface charge state due to the laser irradiation event.

Although charge emission during damage events was expected, even in an atmospheric pressure ambient, the results of measurements on dielectric surfaces were surprising in that a negative surface charge zone more than 10 times the diameter of the damaged site was created.[58] Typical results for a damage site on an HfO_2 thin film are shown in Figure 8.10. This represents a heavily damaged site irradiated by 10 pulses from a Nd:YAG laser. Although the damage spot diameter was less than 0.4 mm, the charged region extends to a diameter of over 6 mm. Furthermore, the charge decayed slowly from the damage region without apparent diffusion. Figure 8.11 shows the charge density decay rate for the same HfO_2 film damage site; the 1/e decay time was longer than one hour. This suggests that neighboring damage test sites might interact over longer distances than had been previously suspected. In addition, greater instrumental sensitivity might allow the extraction of information from predamage irradiation events.

Surface Phase and Structure Analysis

The other measurements desired for postmortem study of laser damage are those which give information about the phase and state of the material at the affected

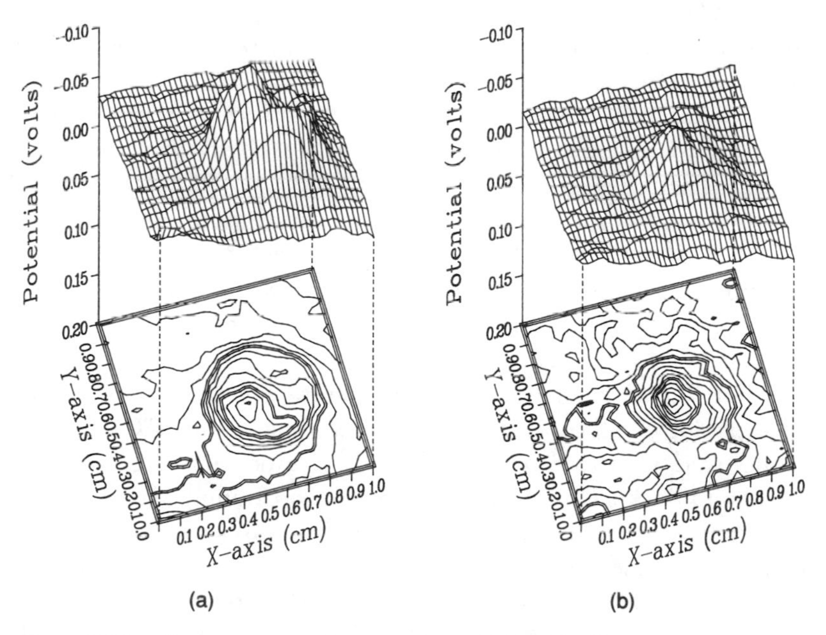

Figure 8.10 Surface potential scans for a HfO_2 thin-film damage site (10 pulses at 50 J/cm^2): (*a*) *t* = 0 and (*b*) *t* = 1 h. (From Becker et al.[58])

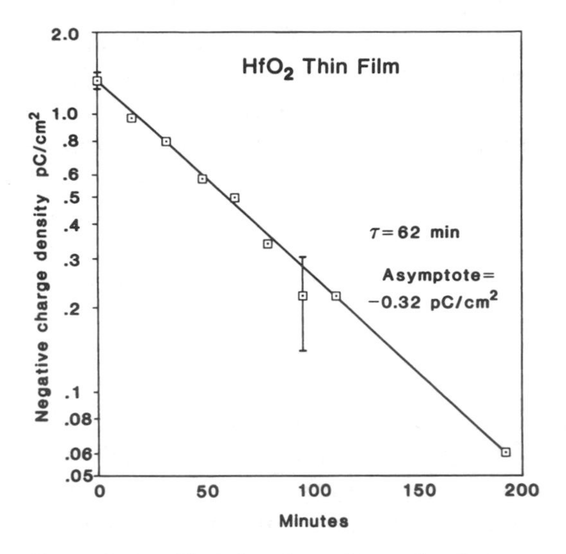

Figure 8.11 **Plot of surface charge density versus time for the same HfO_2 thin-film sample shown in Figure 8.10. (From Reference 58.)**

surface. The most important of these parameters are material crystallinity and crystal phase, stress state, and, for thin films, change in thickness. Traditional methods for such measurements often employ X-ray techniques. Unfortunately, X-ray techniques have inadequate transverse spatial resolution and lose their sensitivity for thin surface layers and thin films. The most successful technique to date for making these measurements at laser damage sites is Raman spectroscopy and the Raman microprobe.

Raman diagnostic techniques have been employed with success in the study of TiO_2 thin films for which three phases are commonly observed: amorphous, anatase, and rutile. The use of Raman spectroscopy to study this system was a natural outgrowth of Raman spectroscopic studies of the as-grown and annealed states of TiO_2 films.[59] With the availability of Raman instruments able to discriminate against the substrate signal, this technique was used to study the time resolved changes in the Raman spectra of laser-irradiated sites[60] and to investigate the structure of very small areas using the Raman microprobe.[61]

The capability of Raman techniques to contribute to the characterization of laser-damage morphologies is best illustrated by a recent study on TiO_2 films.[62] Raman techniques were used to determine the local composition, local residual stress, and relative film thickness at various positions in and around a laser-damage site. An example is given for a Nd:YAG laser damaged ion-beam sputtered TiO_2 film site. Raman probing and optical microscopy identified the annular structure of phases shown in Figure 8.12. A microprobe scan of radial positions through the mixture and anatase phases was taken and peak shifts were noted. A plot of peak shift, correlated to residual stress, versus radial position is shown in Figure 8.13.

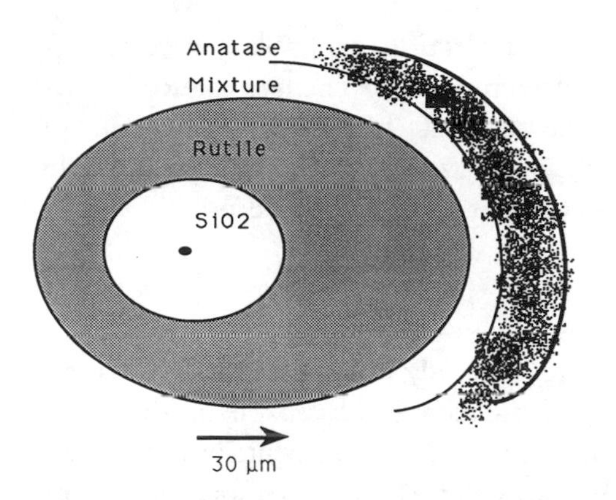

Figure 8.12 Spatial map of the phases at a laser-damage site on an ion-beam sputtered TiO_2 thin film on fused silica determined using the Raman microprobe. (From White et al.[62])

The zero point is approximately at the outer edge of the rutile phase, not at the damage-spot center. It was concluded that the residual stress of the anatase phase increased radially out to the value associated with the expected high compressive stress of the undamaged film.

Raman techniques have clearly demonstrated a special usefulness for characterizing laser damage, but it is not the ultimate technique for all materials. Several new techniques are worth mentioning. For highly perfect material surfaces where defects do not play a dominant role in laser damage, electron diffraction techniques

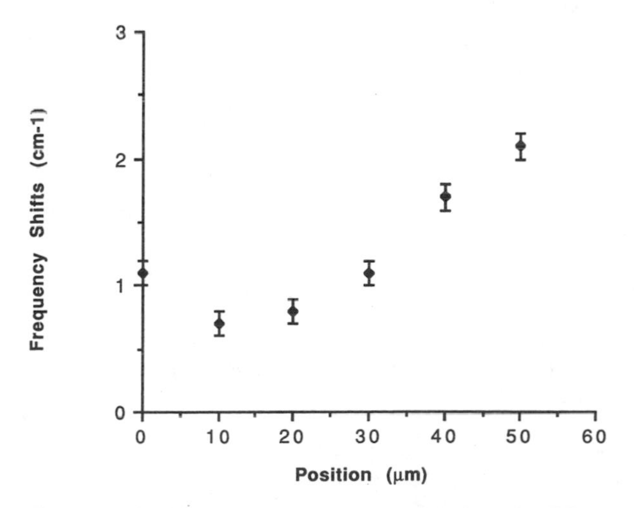

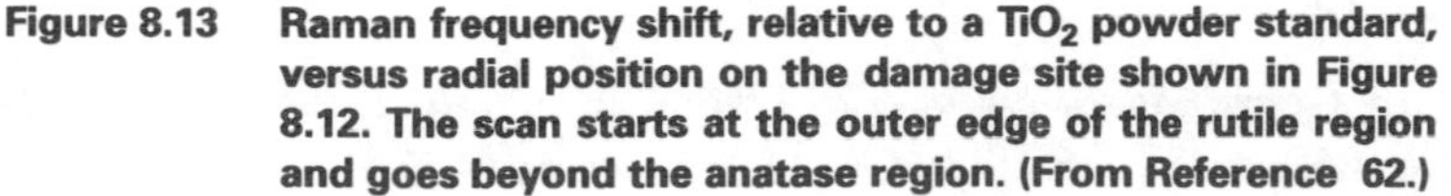

Figure 8.13 Raman frequency shift, relative to a TiO_2 powder standard, versus radial position on the damage site shown in Figure 8.12. The scan starts at the outer edge of the rutile region and goes beyond the anatase region. (From Reference 62.)

(low-energy electron diffraction [LEED] and electron energy-loss spectroscopy [EELS]) have been shown to be useful in detecting subtle changes induced by the laser beam.[63] A second promising technique has yet to be applied to laser damage; it is the use of picosecond laser-generated acoustic pulses to characterize thin-film thicknesses, interface bonds, and imperfections at a locally probed point.[64]

8.5 Future Directions

Future application of material diagnostics to the study of laser-induced damage in optical materials holds special promise for those techniques capable of identifying sites that will damage or of identifying the mechanisms that contribute to multi-pulse processes such as conditioning. In particular, the use of transient thermal imaging techniques to identify predamage sites could be very valuable since it is an anomalous local temperature increase that leads to damage at one particular position within an irradiation spot rather than another. The degree of temperature rise could perhaps be correlated with a temperature rise for damage, so that a non-destructive measure of damage onset might be achievable by this technique. PDT techniques may also have a role to play in damage onset prediction for the same reason: it allows local temperature measurement during a laser pulse and hence damage onset prediction.

To understand multipulse effects, researchers need further understanding of the pulse-to-pulse material changes. The further study of particle emission, stress relief, and postmortem material state should lead to progress in this area. However, since conditioning is dependent on the process used to create a particular surface or film, the diagnosis of conditioning will proceed in parallel with improvements in film-deposition and surface-finishing techniques. Finally, there is a clear need for new materials characterization techniques to be applied to understand the accumulation effect in laser damage of semiconductors.

References

1 M. D. Crisp, N. L. Boling, and G. Dube. *Appl. Phys. Lett.* **21**, 364–366, 1972.

2 A. F. Stewart and A. H. Guenther. In *Laser Induced Damage in Optical Materials: 1987.* NIST Spec. Pub. No. 756. NIST, Boulder, CO, 1988, pp. 142–150.

3 S. Penzoldt, A. P. Elg, M. Reichling, J. Reif, and E. Matthias. *Appl. Phys. Lett.* **53**, 2005–2007, 1988.

4 S. Penzoldt, A. P. Elg, J. Reif, and E. Matthias. In *Laser Induced Damage in Optical Materials: 1989.* NIST Spec. Pub. No. 801. NIST, Boulder, CO, 1990, pp. 180–186.

5 S. R. Foltyn. In *Laser Induced Damage in Optical Materials: 1982.* NIST Spec. Pub. No. 669. NIST, Boulder, CO, 1983, pp. 368–374.

6 S. C. Seitel and J. O. Porteus. In *Laser Induced Damage in Optical Materials: 1983.* NBS Spec. Pub. No. 688. NBS, Boulder, CO, 1985, pp. 502–512.

7 J. O. Porteus and S. C. Seitel. *Applied Optics.* **23**, 3796–3805, 1984.

8 R. M. O'Connell. In *Laser Induced Damage in Optical Materials: 1990.* SPIE Vol. 1441. SPIE, Bellingham, WA, 1991, pp. 406–419.

9 C. D. Marrs, J. O. Porteus, and J. R. Palmer. In *Laser Induced Damage in Optical Materials: 1983.* NBS Spec. Pub. No. 688. NBS, Boulder, CO, 1985, pp. 378–384.

10 C. D. Marrs, J. O. Porteus, and J. R. Palmer. *J. Appl. Phys.* **57**, 1719–1722, 1985.

11 C. D. Marrs, S. J. Walker, M. B. Moran, and J. O. Porteus. In *Laser Induced Damage in Optical Materials: 1986.* NIST Spec. Pub. No. 752. NIST, Boulder, CO, 1987, pp. 245–250.

12 S. E. Clark and D. C. Emmony. In *Laser Induced Damage in Optical Materials: 1988.* NIST Spec. Pub. No. 775. NIST, Boulder, CO, 1989, pp. 62–72.

13 C. R. Wolfe, M. R. Kozlowski, J. H. Campbell, F. Rainer, A. J. Morgan, and R. P. Gonzalez. In *Laser Induced Damage in Optical Materials: 1989.* NIST Spec. Pub. No. 801. NIST, Boulder, CO, 1990, pp. 360–375.

14 M. R. Kozlowski, C. R. Wolfe, M. C. Staggs, and J. H. Campbell. In *Laser Induced Damage in Optical Materials: 1989.* NIST Spec. Pub. No. 801. NIST, Boulder, CO, 1990, pp. 376–390.

15 N. C. Kerr and D. C. Emmony. In *Laser Induced Damage in Optical Materials: 1989.* NIST Spec. Pub. No. 801. NIST, Boulder, CO, 1990, pp. 164–179.

16 M. R. Kozlowski, M. C. Staggs, F. Rainer, and J. H. Stathis. In *Laser Induced Damage in Optical Materials: 1990.* SPIE Vol. 1441. SPIE, Bellingham, WA, 1991, pp. 269–282.

17 M. Schildbach, L. L. Chase, and A. V. Hamza. In *Laser Induced Damage in Optical Materials: 1990.* SPIE Vol. 1441. SPIE, Bellingham, WA, 1991, pp. 287–293.

18 J. W. Arenberg and M. E. Frink. In *Laser Induced Damage in Optical Materials: 1987.* NIST Spec. Pub. No. 756. NIST, Boulder, CO, 1988, pp. 430–439.

19 J. W. Arenberg and D. W. Mordant. In *Laser Induced Damage in Optical Materials: 1988.* NIST Spec. Pub. No. 775. NIST, Boulder, CO, 1989, pp. 516–519.

20 P. A. Temple, D. Milam, and H. Lowdermilk. In *Laser Induced Damage in Optical Materials: 1979.* NBS Spec. Pub. No. 568. NBS, Boulder, CO, 1980, pp. 229–236.

21 P. A. Temple, W. H. Lowdermilk, and D. Milam. *Appl. Opt.* **21**, 3249–3255, 1982.

22 A. J. Weber, A. F. Stewart, G. J. Exarhos, and W. K. Stowell. In *Laser Induced Damage in Optical Materials: 1986.* NIST Spec. Pub. No. 752. NIST, Boulder, CO, 1987, pp. 542–556.

23 A. F. Stewart and A. H. Guenther. In *Laser Induced Damage in Optical Materials: 1988.* NIST Spec. Pub. No. 775. NIST, Boulder, CO, 1989, pp. 176–182.

24 A. F. Stewart, A. H. Guenther, and F. E. Domann. In *Laser Induced Damage in Optical Materials: 1987.* NIST Spec. Pub. No. 756. NIST, Boulder, CO, 1988, pp. 369–387.

25 S. P. Fry, R. M. Walser, and M. F. Becker. In *Laser Induced Damage in Optical Materials: 1987.* NIST Spec. Pub. No. 756. NIST, Boulder, CO, 1988, pp. 492–500.

26 H. H. Musal. In *Laser Induced Damage in Optical Materials: 1979.* NBS Spec. Pub. No. 568. NBS, Boulder, CO, 1980, pp. 159–173.

27 N. Koumvakalis, C. S. Lee, and M. Bass. *Opt. Eng.* **22**, 419–423, 1983.

28 C. S. Lee, N. Koumvakalis, and M. Bass. *J. Appl. Phys.* **54**, 5727–5731, 1983.

29 Y. Jee, M. F. Becker, and R. M. Walser. *J. Opt. Soc. Am. B.* **5**, 648–658, 1988.

30 M. F. Becker, C. Ma, and R. M. Walser. *Appl. Opt.* **30**, 5239–5246, 1991.

31 Y. K. Jhee, M. F. Becker, and R. M. Walser. *J. Opt. Soc. Am. B.* **2**, 1626–1633, 1985.

32 P. M. Fauchet and A. E. Siegman. In *Laser Induced Damage in Optical Materials: 1984.* NBS Spec. Pub. No. 727. NBS, Boulder, CO, 1985, pp. 147–153.

33 J. R. Platenack, R. M. Walser, and M. F. Becker. In *Laser Induced Damage in Optical Materials: 1986.* NIST Spec. Pub. No. 752. NIST, Boulder, CO, 1987, pp. 216–231.

34 R. Swimm and G. Wiemokly. In *Laser Induced Damage in Optical Materials: 1989.* NIST Spec. Pub. No. 801. NIST, Boulder, CO, 1990, pp. 291–298.

35 R. T. Swimm. In *Laser Induced Damage in Optical Materials: 1990.* SPIE Vol. 1441. SPIE, Bellingham, WA, 1991, pp. 45–55.

36 E. Matthias, H. Gronbeck, E. Hunger, J. Jauregui, H. Pietsch, M. Reichling, E. Welsch, and Z. L. Wu. "Frequency- and Time-Resolved Photothermal Investigations of Thin Films." In *Photoacoustic and Photothermal Phenomena III.* Proceedings of the 7th International Topical Meeting: 1991. Springer-Verlag, Berlin, 1992.

37 Z. L. Wu et al. In *Laser Induced Damage in Optical Materials: 1991.* SPIE Vol. 1624. SPIE, Bellingham, WA, 1992, pp. 13–24, 271–281, 331–345, 386–396.

38 R. T. Bailey, F. R. Cruickshank, D. Pugh, and A. McLeod. *Proceedings.* SPIE Vol. 369. SPIE, Bellingham, WA, 1982, pp. 88–89.

39 A. F. Stewart, A. Rusek, and A. H. Guenther. In *Laser Induced Damage in Optical Materials: 1988.* NIST Spec. Pub. No. 775. NIST, Boulder, CO, 1989, pp. 245–258.

40 M. A. Olmstead, N. M. Amer, S. Kohn, D. Fournier, and A. C. Boccara. *Appl. Phys. A.* **32**, 141–154, 1983.

41 C. Karner, A. Mandel, and F. Träger. *Appl. Phys. A.* **38**, 19–21, 1985.

42 J. Opsal, A. Rosencwaig, and D. L. Willenborg. *Appl. Opt.* **22**, 3169–3176, 1983.

43 R. W. Dryfus, F. A. McDonald, and R. J. von Gutfeld. *Appl. Phys. Lett.* **50**, 1491–1493, 1987.

44 R. J. von Gutfeld, F. A. McDonald, and R. W. Dryfus. *Appl. Phys. Lett.* **49**, 1059–1061, 1986.

45 *Photoacoustic, Photothermal and Photochemical Processes at Surfaces and in Thin Films.* (P. Hess, Ed.) Topics in Current Physics, Vol. 47. Springer-Verlag, Berlin, 1989.

46 M. F. Becker, C. Ma, and R. M. Walser. In *Laser Induced Damage in Optical Materials: 1990.* SPIE Vol. 1441. SPIE, Bellingham, WA, 1991, pp. 174–187.

47 J. Jauregui, Z. L. Wu, and E. Matthias. "Pulsed Laser-Induced Temperature Changes Measured by Photothermal Deformation Technique." In *Photoacoustic and Photothermal Phenomena III.* Proceedings of the 7th International Topical Meeting, 1991. Springer-Verlag, Berlin, 1992.

48 Z. L. Wu, J. F. Tang, and B. X. Shi. In *Laser Induced Damage in Optical Materials: 1988.* NIST Spec. Pub. No. 775. NIST, Boulder, CO, 1989, pp. 348–355.

49 Z. L. Wu, M. Reichling, Z. X. Fan, and Z. J. Wang. In *Laser Induced Damage in Optical Materials: 1990.* SPIE Vol. 1441. SPIE, Bellingham, WA, 1991, pp. 214–227.

50 Z. L. Wu, M. Reichling, Z. X. Fan, and Z. J. Wang. In *Laser Induced Damage in Optical Materials: 1990.* SPIE Vol. 1441. SPIE, Bellingham, WA, 1991, pp. 200–213.

51 Y. K. Jhee, M. F. Becker, and R. M. Walser. *J. Opt. Soc. Am B.* **2**, 1626–1633, 1985.

52 J. A. Kardach, A. F. Stewart, and A. H. Guenther. In *Laser Induced Damage in Optical Materials: 1986.* NIST Spec. Pub. No. 752. NIST, Boulder, CO, 1987, pp. 488–504.

53 W. J. Siekhaus, L. L. Chase, and D. Milam. In *Laser Induced Damage in Optical Materials: 1985.* NBS Spec. Pub. No. 746. NBS, Boulder, CO, 1986, pp. 509–514.

54 L. L. Chase and L. K. Smith. In *Laser Induced Damage in Optical Materials: 1987.* NIST Spec. Pub. No. 756. NIST, Boulder, CO, 1988, pp. 165–174.

55 F. E. Domann, A. F. Stewart, and A. H. Guenther. In *Laser Induced Damage in Optical Materials: 1987.* NIST Spec. Pub. No. 756. NIST, Boulder, CO, 1988, pp. 175–186.

56 H. F. Arlinghaus, W. F. Calaway, D. M. Gruen, and L. L. Chase. In *Laser Induced Damage in Optical Materials: 1988.* NIST Spec. Pub. No. 775. NIST, Boulder, CO, 1989, pp. 140–151.

57 M. F. Becker, J. A. Kardach, A. F. Stewart, and A. H. Guenther. In *Laser Induced Damage in Optical Materials: 1984.* NBS Spec. Pub. No. 727. NBS, Boulder, CO, 1985, pp. 116–126.

58 M. F. Becker, A. F. Stewart, J. A. Kardach, and A. H. Guenther. *Appl. Opt.* **26**, 805–812, 1987.

59 L. S. Hsu, R. Solanki, G. J. Collins, and C. Y. She. *Phys. Lett.* **45**, 1065–1067, 1984.

60 G. J. Exarhos and P. L. Morse. In *Laser Induced Damage in Optical Materials: 1984.* NBS Spec. Pub. No. 727. NBS, Boulder, CO, 1985, pp. 262–271.

61 D. M. Freidrich and G. J. Exarhos. In *Laser Induced Damage in Optical Materials: 1985.* NBS Spec. Pub. No. 746. NBS, Boulder, CO, 1986, pp. 374–382.

62 P. L. White, G. J. Exarhos, M. Bowden, N. M. Dixon, and D. J. Gardiner. *J. Mater. Res.* **6**, 126–133, 1991.

63 M. A. Schildbach and A. V. Hamza. In *Laser Induced Damage in Optical Materials: 1990.* SPIE Vol. 1441. SPIE, Bellingham, WA, 1991, pp. 139–145.

64 C. Thomsen, H. J. Maris, and J. Tauc. *Thin Solid Films.* **154**, 217–223, 1987.

Appendix: Technique Summaries

The technique summaries on subsequent pages of this appendix marked with an asterisk are reprinted from the lead volume of this series, *Encyclopedia of Materials Characterization*, by C. Richard Brundle, Charles A. Evans, Jr., and Shaun Wilson; they are summaries of full-length articles appearing there. The list below organizes all the techniques in this appendix into related groups and gives their appendix order number.

Physical Properties

- Ellipsometry
  - Variable-angle spectroscopic ellipsometry (VASE) **21**
- Optical Spectroscopy
  - Luminescence, fluorescence **10**
  - Modulation spectroscopy (surface charge density mapping) **7**

Surface/Interface Morphology

- Optical microscopy **6**
  - Nomarski (differential interference contrast)
  - Total internal reflection microscopy **19**
- Photothermal displacement technique **11**
- Profilometry measurements for the determination of surface smoothness **18**
- Total integrated and angle-resolved scatterometry (TIS, ARS) **9**
- Scanning tunneling microscopy (STM), atomic force microscopy (AFM) **16**

Microstructure/Atomic Positions

- X-ray diffraction **22**
- Electron microscopy and diffraction
  - Cross-section transmission electron microscopy (XTEM) **20**
  - Scanning transmission electron microscopy (STEM) **15**
  - Scanning electron microscopy (SEM) **14**
- Electron energy-loss spectroscopy (EELS) **3**

Chemical Bonding Specific Spectroscopies

- Raman spectroscopy **12**
  - In situ Raman methods
  - Raman microprobe spectroscopy
  - Time-resolved Raman spectroscopy
- Fourier transform infrared absorption spectroscopy **5**

Atom Specific Spectroscopies

- Auger electron spectroscopy (AES) **1**
- Static secondary ion mass spectometry (Static SIMS) **17**
- Cathodoluminescence spectroscopy **2**
- X-ray photoelectron spectroscopy (XPS) **24**
- Rutherford backscattering spectroscopy (RBS) **13**
- Nuclear reaction analysis (NRA) **8**
- Energy-dispersive X-ray spectroscopy (EDS) **4**
- X-ray fluorescence (XRF) **23**

Auger Electron Spectroscopy (AES)* 1

Auger electron spectroscopy (AES) uses a focused electron beam to create secondary electrons near the surface of a solid sample. Some of these (the Auger electrons) have energies characteristic of the elements and, in many cases, of the chemical bonding of the atoms from which they are released. Because of their characteristic energies and the shallow depth from which they escape without energy loss, Auger electrons are able to characterize the elemental composition and, at times, the chemistry of the surfaces of samples. When used in combination with ion sputtering to gradually remove the surface, Auger spectroscopy can similarly characterize the sample in depth. The high spacial resolution of the electron beam and the process allows microanalysis of three-dimensional regions of solid samples. AES has the attributes of high lateral resolution, relatively high sensitivity, standardless semiquantitative analysis, and chemical bonding information in some cases.

Range of elements	All except hydrogen and helium
Destructive	No, except to electron beam-sensitive materials and during depth profiling
Elemental Analysis	Yes, semiquantitative without standards; quantitative with standards
Absolute sensitivity	100 ppm for most elements, depending on the matrix
Chemical state information	Yes, in many materials
Depth probed	5–100 Å
Depth profiling	Yes, in combination with ion-beam sputtering
Lateral resolution	300 Å for Auger analysis
Imaging/mapping	Yes, called scanning Auger microscopy (SAM)
Sample requirements	Vacuum-compatible materials
Main use	Elemental composition of inorganic materials
Instrument cost	$100,000–$800,000
Size	10 ft × 15 ft

Cathodoluminescence (CL)* 2

In cathodoluminescence (CL) analysis, electron-beam bombardment of a solid placed in vacuum causes emission of photons (in the ultraviolet, visible, and near-infrared ranges) due to the recombination of electron–hole pairs generated by the incident energetic electrons. The signal provides a means for CL microscopy (i.e., CL images are displayed on a CRT) and spectroscopy (i.e., luminescence spectra from selected areas of the sample are obtained) analysis of luminescent materials using electron probe instruments. CL microscopy can be used for uniformity characterization (e.g., mapping of defects and impurity segregation studies), whereas CL spectroscopy provides information on various electronic properties of materials.

Range of elements	Not element specific
Chemical bonding information	Sometimes
Nondestructive	Yes; caution—in certain cases electron bombardment may ionize or create defects
Detection limits	In favorable cases, dopant concentrations down to 10^{14} atoms/cm^3
Depth profiling	Yes, by varying the range of electron penetration (between about 10 nm and several μm), which depends on the electron-beam energy (1–40 keV).
Lateral resolution	On the order of 1 μm; down to about 0.1 μm in special cases
Imaging/mapping	Yes
Sample requirements	Solid, vacuum compatible
Quantification	Difficult, standards needed
Main use	Nondestructive qualitative and quantitative analysis of impurities and defects and their distributions in luminescent materials
Instrument cost	$25,000–$250,000
Size	Small add-on item to SEM, TEM

Electron Energy-Loss Spectroscopy in the Transmission Electron Microscope (EELS)* 3

In electron energy-loss spectroscopy (EELS) a nearly monochromatic beam of electrons is directed through an ultrathin specimen, usually in a transmission (TEM) or scanning transmission (STEM) electron microscope. As the electron beam propagates through the specimen, it experiences both elastic and inelastic scattering with the constituent atoms, which modifies its energy distribution. Each atomic species in the analyzed volume causes a characteristic change in the energy of the incident beam; the changes are analyzed by means of a electron spectrometer and counted by a suitable detector system. The intensity of the measured signal can be used to determine quantitatively the local specimen concentration, the electronic and chemical structure, and the nearest neighbor atomic spacings.

Range of elements	Lithium to uranium; hydrogen and helium are sometimes possible
Destructive	No
Chemical bonding information	Yes, in the near-edge structure of edge profiles
Depth profiling capabilities	None, the specimen is already thin
Quantification	Without standards ~±10–20% at.; with standards ~1–2% at.
Detection limits	~10^{-21} g
Depth probed	Thickness of specimen (≤ 2000 Å)
Lateral resolution	1 nm to 10 μm, depending on the diameter of the incident electron probe and the thickness of the specimen
Imaging capabilities	Yes
Sample requirements	Solids; specimens must be transparent to electrons and ~100–2000 Å thick
Main use	Light element spectroscopy for concentration, electronic, and chemical structure analysis at ultrahigh lateral resolution in a TEM or STEM
Cost	As an accessory to a TEM or STEM: $50,000–$150,000 (does not include electron microscope cost)

Energy-Dispersive X-Ray Spectroscopy (EDS)* 4

When the atoms in a material are ionized by a high-energy radiation they emit characteristic X rays. EDS is an acronym describing a technique of X-ray spectroscopy that is based on the collection and energy dispersion of characteristic X rays. An EDS system consists of a source of high-energy radiation, usually electrons; a sample; a solid state detector, usually made from lithium-drifted silicon, Si (Li); and signal processing electronics. EDS spectrometers are most frequently attached to electron column instruments. X rays that enter the Si (Li) detector are converted into signals which can be processed by the electronics into an X-ray energy histogram. This X-ray spectrum consists of a series of peaks representative of the type and relative amount of each element in the sample. The number of counts in each peak may be further converted into elemental weight concentration either by comparison with standards or by standardless calculations.

Range of elements	Boron to uranium
Destructive	No
Chemical bonding information	Not readily available
Quantification	Best with standards, although standardless methods are widely used
Accuracy	Nominally 4–5%, relative, for concentrations > 5% wt.
Detection limits	100–200 ppm for isolated peaks in elements with Z > 11, 1–2% wt. for low-Z and overlapped peaks
Lateral resolution	0.5–1 μm for bulk samples; as small as 1 nm for thin samples in STEM
Depth sampled	0.02 to μm, depending on Z and keV
Imaging/mapping	In SEM, EPMA, and STEM
Sample requirements	Solids, powders, and composites; size limited only by the stage in SEM, EPMA, and XRF; liquids in XRF; 3 mm diameter thin foils in TEM
Main use	To add analytical capability to SEM, EPMA, and TEM
Cost	$25,000–$100,000, depending on accessories (not including the electron microscope)

Fourier Transform Infrared Spectroscopy (FTIR)* 5

The vibrational motions of the chemically bound constituents of matter have frequencies in the infrared regime. The oscillations induced by certain vibrational modes provide a means for matter to couple with an impinging beam of infrared electromagnetic radiation and to exchange energy with it when the frequencies are in resonance. In the infrared experiment, the intensity of a beam of infrared radiation is measured before (I_0) and after (I) it interacts with the sample as a function of light frequency, $\{w_i\}$. A plot of I/I_0 versus frequency is the "infrared spectrum." The identities, surrounding environments, and concentrations of the chemical bonds that are present can be determined.

Information	Vibrational frequencies of chemical bonds
Element range	All, but not element specific
Destructive	No
Chemical bonding information	Yes, identification of functional groups
Depth profiling	No, not under standard conditions
Depth probed	Sample dependent, from μm's to 10 nm
Detection limits	Ranges from undetectable to < 10^{13} bonds/cc. Sub-monolayer sometimes
Quantification	Standards usually needed
Reproducibility	0.1% variation over months
Lateral resolution	0.5 cm to 20 μm
Imaging/mapping	Available, but not routinely used
Sample requirements	Solid, liquid, or gas in all forms; vacuum not required
Main use	Qualitative and quantitative determination of chemical species, both trace and bulk, for solids and thin films. Stress, structural inhomogeneity
Instrument cost	\$50,000–\$150,000 for FTIR; \$20,000 or more for non-FT spectrophotometers
Instrument size	Ranges from desktop to (2 × 2 m)

Light Microscopy* 6

The light microscope uses the visible or near visible portion of the electromagnetic spectrum; light microscopy is the interpretive use of the light microscope. This technique, which is much older than other characterization instruments, can trace its origin to the 17th century. Modern analytical and characterization methods began about 150 years ago when thin sections of rocks and minerals, and the first polished metal and metal-alloy specimens were prepared and viewed with the intention of correlating their structures with their properties. The technique involves, at its very basic level, the simple, direct visual observation of a sample with white-light resolution to 0.2 μm. The morphology, color, opacity, and optical properties are often sufficient to characterize and identify a material.

Range of samples characterized	Almost unlimited for solids and liquid crystals
Destructive	Usually nondestructive; sample preparation may involve material removal
Quantification	Via calibrated eyepiece micrometers and image analysis
Detection limits	To sub-ng
Resolving power	0.2 μm with white light
Imaging capabilities	Yes
Main use	Direct visual observation; preliminary observation for final characterization, or preparative for other instrumentation
Instrument cost	$2,500–$50,000 or more
Size	Pocket to large table

Modulation Spectroscopy* 7

Modulation spectroscopy is a powerful experimental method for measuring the energy of transitions between the filled and empty electronic states in the bulk (band gaps) or at surfaces of semiconductor materials over a wide range of experimental conditions (temperature, ambients, etc.). By taking the derivative of the reflectance (or transmittance) of a material in an analog manner, it produces a series of sharp, derivative-like spectral features corresponding to the photon energy of the transitions. These energies are sensitive to a number of internal and external parameters such as chemical composition, temperature, strain, and electric and magnetic fields. The line widths of these spectral features are a function of the quality of the material.

Destructiveness	Some methods are nondestructive
Depth probed	For bulk applications 0.1–1 μm; for surface applications one monolayer is possible
Lateral resolution	Down to 100 μm
Image/mapping	Yes
Sensitivity	Alloy composition (e.g., $Ga_{1-x}Al_xAs$) $\Delta x = 0.005$; carrier concentration 10^{15}–10^{19} cm^{-3}
Main uses	Contactless, nondestructive monitoring of band gaps in semiconductors; wide range of temperatures and ambients (air, ultrahigh vacuum); in situ monitoring of semiconductor growth
Instrument cost	\$30,000–\$100,000
Size	For most methods about 2 × 3 ft

Nuclear Reaction Analysis (NRA)* 8

In nuclear reaction analysis (NRA), a beam of charged particles with energy from a few hundred keV to several MeV is produced in an accelerator and bombards a sample. Nuclear reactions with low-Z nuclei in the sample are induced by the ion beam. Products of these reactions (typically protons, deuterons, tritons, He, α particles, and γ rays) are detected, producing a spectrum of particle yield versus energy. Depth information is obtained from the spectrum using energy loss rates for incident and product ions traveling through the sample. Particle yields are converted to concentrations with the use of experimental parameters and nuclear reaction cross sections.

Range of elements	Hydrogen to calcium; specific isotopes
Destructive	No, but some materials may be damaged by ion beams
Chemical bonding information	No
Depth profiling	Yes
Quantification	Yes, standards usually unnecessary
Accuracy	A few percent to tens of percent
Detection limits	Varies with specific reaction; typically 10–100 ppm
Depth probed	Several μm
Depth resolution	Varies with specific reaction; typically a few nm to hundreds of nm
Lateral resolution	Down to a few μm with microbeams
Imaging/mapping	Yes, with microbeams
Sample requirements	Solid conductors and insulators
Main use	Quantitative measurement of light elements (particularly hydrogen) in solid materials, without standards; has isotope selectivity
Instrument Cost	Several million dollars for high-energy ion accelerator
Size	Large laboratory for accelerator

Optical Scatterometry* 9

Optical scatterometry involves illuminating a sample with light and measuring the angular distribution of light which is scattered. The technique is useful for characterizing the topology of two general categories of surfaces. First, surfaces that are nominally smooth can be examined to yield the root-mean-squared (rms) roughness and other surface statistics. Second, the shapes of structure (lines) of periodically patterned surfaces can be characterized. The intensity of light diffracted into the various diffraction orders from the periodic structure is indicative of the shape of the lines. If the line shape is influenced by steps involved in processing the sample, the scattering technique can be used to monitor the process. This has been applied to several steps involved in microelectronics processing. Scatterometry is noncontact, nondestructive, fast, and often yields quantitative results. For some applications it can be used in situ.

Parameters measured	Surface topography (rms roughness, rms slope, and power spectrum of structure); scattered light; line shape of periodic structure (width, side wall angle, height, and period)
Destructive	No
Vertical resolution	≥ 0.1 nm
Lateral resolution	≥ λ/2 for topography characterization, much smaller for periodic structure characterization (λ is the laser wavelength used to illuminate the sample)
Main uses	Topography characterization of nominally smooth surfaces; process control when characterizing periodic structure; can be applied in situ in some cases; rapid; amenable to automation
Quantitative	Yes
Mapping capabilities	Yes
Instrument cost	$10,000–$200,000 or more
Size	1 ft × 1 ft to 4 ft × 8 ft

Photoluminescence (PL)* 10

In photoluminescence one measures physical and chemical properties of materials by using photons to induce excited electronic states in the material system and analyzing the optical emission as these states relax. Typically, light is directed onto the sample for excitation, and the emitted luminescence is collected by a lens and passed through an optical spectrometer onto a photodetector. The spectral distribution and time dependence of the emission are related to electronic transition probabilities within the sample, and can be used to provide qualitative and, sometimes, quantitative information about chemical composition, structure (bonding, disorder, interfaces, quantum wells), impurities, kinetic processes, and energy transfer.

Destructiveness	Nondestructive
Depth probed	0.1–3 μm; limited by light penetration depth and carrier diffusion length
Lateral resolution	Down to 1–2 μm
Quantitative abilities	Intensity-based impurity quantification to several percent possible; energy quantification very precise
Sensitivity	Down to parts-per-trillion level, depending on impurity species and host
Imaging/mapping	Yes
Sample requirements	Liquid or solid having optical transitions; probe size 2 μm to a few cm
Main uses	Band gaps of semiconductors; carrier lifetimes; shallow impurity or defect detection; sample quality and structure
Instrument cost	Less than $10,000 to over $200,000
Size	Table top to small room

Photothermal Displacement Technique 11

A solid sample is irradiated by an intensity-modulated laser beam (pump). The sample absorbs a fraction of the incident energy, which is transformed to heat in the near-surface layer. The resulting thermal wave induces a transient deformation on the sample surface, which is monitored by a second, lower-energy laser beam (probe) that is imaged onto the sample surface. The probe beam is deflected by the surface deformation and detected by a position-sensitive detector. Measurements correlate directly with the slope of the surface deformation and provide information concerning both optical and thermal properties of the sample. Localized surface defects can be probed by spatially scanning probe and pump beams over the surface of interest.

Optical and thermal defects can be distinguished from one another in thin films by varying the modulation frequency. At relatively low frequencies, the thermal diffusion length is substantially larger than the film thickness and the resulting signal is sensitive only to optical non-uniformities in the film. At high modulation frequencies, where the thermal diffusion length is comparable or smaller than the film thickness, the measured signal contains a marked thermal contribution from the film as well. Therefore, spatially resolved measurements acquired at low and high modulation frequencies will enable both thermal and optical inhomogeneities to be imaged.

To perform these measurements, an intensity-modulated pump laser (10 Hz to 10 MHz), such as a 200 mW argon ion laser operating at 514.5 nm, and a 2 mW CW He–Ne laser are imaged (1 μm spot size) onto the sample by means of a microscope objective. The sample is mounted on an x–y translating stage; the reflected probe light is detected by a position-sensitive detector. The absorption sensitivity of this instrument is in the parts-per-million range.

Range of samples characterized	Both transparent and opaque samples can be studied in situ
Destructive	Nondestructive measurement
Quantification	Via image analysis and comparison to reference standards
Detection limits	Can be used to investigate films or surface layers having submicrometer thicknesses
Resolving power	Approximately 1 μm using visible light
Imaging capabilities	Yes

Main use	Characterization of localized optical and thermal inhomogeneities in thin films, identification of subsurface defects, and defects at buried interfaces; useful in the identification of pre-laser damage sites in optical coatings for high energy pulsed laser applications and for measurements of thermal diffusivity in thin films
Instrument cost	$20,000 to $50,000 or more
Size	Small optical table (vibrationally isolated)

Recommended Reading

Jackson, W. B., N. M. Amer, A. C. Boccara, and D. Fournier. "Photothermal Deflection Spectroscopy and Detection," *Appl. Opt.* **20**, 1333, 1981.

Karner, C., A. Mandel, and F. Traeger. "Pulsed Laser Photothermal Displacement Spectroscopy for Surface Studies," *Appl. Phys. A.* **38**, 19, 1985.

Olmstead, M. A., N. M. Amer, S. Kohn, D. Fournier, and A. C. Boccara. "Photothermal Displacement Spectroscopy," *Appl. Phys. A.* **32**, 141, 1983.

Wu, Z. L., M. Reichling, E. Welsch, D. Schaefer, Z. X. Fan, and E. Matthias. "Defect Characterization for Thin Films Through Thermal Wave Detection," in *Laser-Induced Damage in Optical Materials: 1991*, Vol. 1624. (H. E. Bennett, L. L. Chase, A. H. Guenther, B. E. Newnam, and M. J. Soileau, Eds.) Soc. Photo-Opt. Instru. Eng., Bellingham, WA, 1992, p. 331.

Wu, Z. L., Z. J. Wu, Z. X. Fan, and Z. J. Wang. "Photothermal Deflection Microscopy of Optical Coatings," in *Optical Coatings.* (J. F. Tang, Ed.) Shanghai, 1989, p. 226.

Raman Spectroscopy* 12

Raman spectroscopy is the measurement, as a function of wavenumber, of the inelastic light scattering that results from the excitation of vibrations in molecular and crystalline materials. The excitation source is a single line of a continuous gas laser, which permits optical microscope optics to be used for measurement of samples down to a few mm. Raman spectroscopy is sensitive to molecular and crystal structure; applications include chemical fingerprinting, examination of single grains in ceramics and rocks, single-crystal measurements, speciation of aqueous solutions, identification of compounds in bubbles and fluid inclusions, investigations of structure and strain states in polycrystalline ceramics, glasses, fibers, gels, and thin and thick films.

Information	Vibrational frequencies of chemical bonds
Element range	All, but not element specific
Destructive	No, unless sample is susceptible to laser damage
Lateral resolution	1 μm with microfocus instruments
Depth profiling	Limited to transparent materials
Depth probed	Few μm to mm, depending on material
Detection limits	1000 Å normally, submonolayer in special cases
Quantitative	With difficulty; usually qualitative only
Imaging	Usually no, although imaging instruments have been built
Sample requirements	Very flexible: liquids, gases, crystals, polycrystalline solids, powders, and thin films
Main use	Identification of unknown compounds in solutions, liquids, and crystalline materials; characterization of structural order, and phase transitions
Instrument cost	$150,000–$250,000
Size	1.5 m × 2.5 m

Rutherford Backscattering Spectrometry (RBS)* 13

Rutherford backscattering spectrometry (RBS) analysis is performed by bombarding a sample target with a monoenergetic beam of high-energy particles, typically helium, with an energy of a few MeV. A fraction of the incident atoms scatter backwards from heavier atoms in the near-surface region of the target material, and usually are detected with a solid state detector that measures their energy. The energy of a backscattered particle is related to the depth and mass of the target atom, while the number of backscattered particles detected from any given element is proportional to concentration. This relationship is used to generate a quantitative depth profile of the upper 1–2 μm of the sample. Alignment of the ion beam with the crystallographic axes of a sample permits crystal damage and lattice locations of impurities to be quantitatively measured and depth profiled. The primary applications of RBS are the quantitative depth profiling of thin-film structures, crystallinity, dopants, and impurities.

Range of elements	Lithium to uranium
Destructive	~10^{13} He atoms implanted; radiation damage.
Chemical bonding information	No
Quantification	Yes, standardless; accuracy 5–20%
Detection limits	10^{12}–10^{16} atoms/cm^2; 1–10 at. % for low-Z elements; 0–100 ppm for high-Z elements
Lateral resolution	1–4 mm, 1 μm in specialized equipment
Depth profiling	Yes and nondestructive
Depth resolution	2–30 nm
Maximum depth	~2 μm, 20 μm with H^+
Imaging/mapping	Under development
Sample requirements	Solid, vacuum compatible
Main use	Nondestructive depth profiling of thin films, crystal damage information
Instrument cost	$450,000–$1,000,000
Size	2 m × 7 m

Scanning Electron Microscopy (SEM)* 14

The scanning electron microscope (SEM) is often the first analytical instrument used when a "quick look" at a material is required and the light microscope no longer provides adequate resolution. In the SEM an electron beam is focused into a fine probe and subsequently raster scanned over a small rectangular area. As the beam interacts with the sample it creates various signals (secondary electrons, internal currents, photon emission, etc.), all of which can be appropriately detected. These signals are highly localized to the area directly under the beam. By using these signals to modulate the brightness of a cathode ray tube, which is raster scanned in synchronism with the electron beam, an image is formed on the screen. This image is highly magnified and usually has the "look" of a traditional microscopic image but with a much greater depth of field. With ancillary detectors, the instrument is capable of elemental analysis.

Main use	High magnification imaging and composition (elemental) mapping
Destructive	No, some electron beam damage
Magnification range	10×–300,000×; 5000×–100,000× is the typical operating range
Beam energy range	500 eV to 50 keV; typically, 20–30 keV
Sample requirements	Minimal, occasionally must be coated with a conducting film; must be vacuum compatible
Sample size	Less than 0.1 mm, up to 10 cm or more
Lateral resolution	1–50 nm in secondary electron mode
Depth sampled	Varies from a few nm to a few µm, depending upon the accelerating voltage and the mode of analysis
Bonding information	No
Depth profiling capabilities	Only indirect
Instrument cost	$100,000–$300,000 is typical
Size	Electronics console 3 ft × 5 ft; electron beam column 3 ft × 3 ft

Scanning Transmission Electron Microscopy (STEM)* 15

In scanning transmission electron microscopy (STEM) a solid specimen, 5 to 500 nm thick, is bombarded in vacuum by a beam (0.3–50 nm in diameter) of monoenergetic electrons. STEM images are formed by scanning this beam in a raster across the specimen and collecting the transmitted or scattered electrons. Compared to the TEM an advantage of the STEM is that many signals may be collected simultaneously: bright- and dark-field images; convergent beam electron diffraction (CBED) patterns for structure analysis; and energy-dispersive X-ray spectrometry (EDS) and electron energy-loss spectrometry (EELS) signals for compositional analysis. Taken together, these analysis techniques are termed analytical electron microscopy (AEM). STEM provides about 100 times better spatial resolution of analysis than conventional TEM. When electrons scattered into high angles are collected, extremely high-resolution images of atomic planes and even individual heavy atoms may be obtained.

Range of elements	Lithium to uranium
Destructive	Yes, during specimen preparation
Chemical bonding information	Sometimes, from EELS
Quantification	Quantitative compositional analysis from EDS or EELS, and crystal structure analysis from CBED
Accuracy	5–10% relative for EDS and EELS
Detection limits	0.1–3.0% wt. for EDS and EELS
Lateral resolution	Imaging, 0.2–10 nm; EELS, 0.5–10 nm; EDS, 3–30 nm
Imaging/mapping capabilities	Yes, lateral resolution down to < 5 nm
Sample requirements	Solid conductors and coated insulators typically 3 mm in diameter and < 200 nm thick at the analysis point for imaging and EDS, but < 50 nm thick for EELS
Main uses	Microstructural, crystallographic, and compositional analysis; high spatial resolution with good elemental detection and accuracy; unique structural analysis with CBED
Instrument cost	\$500,000–\$2,000,000
Size	3 m × 4 m × 3 m

Scanning Tunneling Microscopy and Scanning Force Microscopy (STM and SFM)* 16

In scanning tunneling microscopy (STM) or scanning force microscopy (SFM), a solid specimen in air, liquid or vacuum is scanned by a sharp tip located within a few Å of the surface. In STM, a quantum-mechanical tunneling current flows between atoms on the surface and those on the tip. In SFM, also known as atomic force microscopy (AFM), interatomic forces between the atoms on the surface and those on the tip cause the deflection of a microfabricated cantilever. Because the magnitude of the tunneling current or cantilever deflection depends strongly upon the separation between the surface and tip atoms, they can be used to map out surface topography with atomic resolution in all three dimensions. The tunneling current in STM is also a function of local electronic structure so that atomic-scale spectroscopy is possible. Both STM and SFM are unsurpassed as high-resolution, three-dimensional profilometers.

Parameters measured	Surface topography (SFM and STM); local electronic structure (STM)
Destructive	No
Vertical resolution	STM, 0.01 Å; SFM, 0.1 Å
Lateral resolution	STM, atomic; SFM, atomic to 1 nm
Quantification	Yes; three-dimensional
Accuracy	Better than 10% in distance
Imaging/mapping	Yes
Field of view	From atoms to > 250 μm
Sample requirements	STM—solid conductors and semiconductors, conductive coating required for insulators; SFM—solid conductors, semiconductors and insulators
Main uses	Real-space three-dimensional imaging in air, vacuum, or solution with unsurpassed resolution; high-resolution profilometry; imaging of nonconductors (SFM).
Instrument cost	$65,000 (ambient) to $200,000 (ultrahigh vacuum)
Size	Table-top (ambient), 2.27–12 in. bolt-on flange (ultrahigh vacuum)

Static Secondary Ion Mass Spectrometry (Static SIMS)* 17

Static secondary ion mass spectrometry (SIMS) involves the bombardment of a sample with an energetic (typically 1–10 keV) beam of particles, which may be either ions or neutrals. As a result of the interaction of these primary particles with the sample, species are ejected that have become ionized. These ejected species, known as secondary ions, are the analytical signal in SIMS.

In static SIMS, the use of a low dose of incident particles (typically less than 5×10^{12} atoms/cm^2) is critical to maintain the chemical integrity of the sample surface during analysis. A mass spectrometer sorts the secondary ions with respect to their specific charge-to-mass ratio, thereby providing a mass spectrum composed of fragment ions of the various functional groups or compounds on the sample surface. The interpretation of these characteristic fragmentation patterns results in a chemical analysis of the outer few monolayers. The ability to obtain surface chemical information is the key feature distinguishing static SIMS from dynamic SIMS, which profiles rapidly into the sample, destroying the chemical integrity of the sample.

Range of elements	Hydrogen to uranium; all isotopes
Destructive	Yes, if sputtered long enough
Chemical bonding information	Yes
Depth probed	Outer 1 or 2 monolayers
Lateral resolution	Down to ~100 μm
Imaging/mapping	Yes
Quantification	Possible with appropriate standards
Mass range	Typically, up to 1000 amu (quadrupole), or up to 10,000 amu (time of flight)
Sample requirements	Solids, liquids (dispersed or evaporated on a substrate), or powders; must be vacuum compatible
Main use	Surface chemical analysis, particularly organics, polymers
Instrument cost	\$500,000–\$750,000
Size	4 ft × 8 ft

Surface Roughness: Measurement, Formation by Sputtering, Impact on Depth Profiling* 18

Surface roughness is commonly measured using mechanical and optical profilers, scanning electron microscopes, and atomic force and scanning tunneling microscopes. Angle-resolved scatterometers can also be applied to this measurement. The analysis surface can be roughened by ion bombardment, and roughness will degrade depth resolution in a depth profile. Rotation of the sample during sputtering can reduce this roughening.

Mechanical Profiler

Depth resolution	0.5 nm
Minimum step	2.5–5 nm
Maximum step	~150 μm
Lateral resolution	0.1–25 μm, depending on stylus radius
Maximum sample size	15-mm thickness, 200-mm diameter
Instrument cost	$30,000–$70,000

Optical Profiler

Depth resolution	0.1 nm
Minimum step	0.3 nm
Maximum step	15 μm
Lateral resolution	0.35–9 μm, depending on optical system
Maximum sample size	125-mm thickness, 100-mm diameter
Instrument cost	$80,000–$100,000

SEM (see SEM summary 14)

Scanning Force Microscope (see STM/SFM summary 16)

Depth resolution	0.01 nm
Lateral resolution	0.1 nm
Instrument cost	$75,000–$150,000

Scanning Tunneling Microscope (see STM/SFM summary 16)

Depth resolution	0.001 μm
Lateral resolution	0.1 nm
Instrument cost	$75,000–$150,000

Optical Scatterometer (see Optical Scatterometry summary 9)

Depth resolution	0.1 nm (root mean square)
Instrument cost	$50,000–$150,000

Total Internal Reflection Microscopy 19

This technique is effective for characterizing subwavelength features on transparent surfaces. Scattered light from surface and near-surface imperfections can be detected readily by means of an optical microscope focused on the sample surface. The instrument can be modified for installation in a vacuum chamber for real-time analysis of film surfaces during deposition. (See Williams et al.)

The sample surface under study is optically coupled to the hypotenuse face of a right-angle isosceles prism by means of a thin layer of index matching fluid. A probe laser beam is directed into the prism by means of an adjustable mirror so that the entering beam is normal to the prism face and incident on the top surface of the optic at an angle slightly greater than the critical angle, θ_c, for total internal reflection. Under these conditions, no specular beam in the region above the sample is seen. However, light scattered by surface and subsurface defects can escape from the optic and be observed in an optical microscope focused on the surface from above.

Range of samples characterized	Method requires transparent samples
Destructive	Usually nondestructive; the index matching fluid can be removed readily
Quantification	Via calculated eyepiece micrometers and image analysis
Detection limits	To sub-ng
Resolving power	0.2 μm with white light
Imaging capabilities	Yes
Main use	Direct visual observation of surface scratches, pits, and subsurface microcracks; preliminary observation for final characterization
Instrument cost	$3000 to $50,000 or more
Size	Pocket to large table

Recommended Reading

Jabr, S. N. "Total Internal Reflection Microscopy: Inspection of Surfaces of High Bulk Scatter Materials," *Appl. Opt.* **24**, 1689, 1985.

Temple, P. A. "Total Internal Reflection Microscopy: A Surface Inspection Technique," *Appl. Opt.* **20**, 2656, 1981.

Williams, F. L., G. A. Peterson, Jr., R. A. Schmel, and C. K. Carniglia. "Observation and Control of Thin-Film Defects Using In-Situ Total Internal Refelction Microscopy," in *Laser Induced Damage in Optical Materials: 1991*, Vol. 1624. (H. E. Bennett, L. L. Chase, A. H. Guenther, B. E. Newnam, and M. J. Soileau, Eds.) Soc. Photo-Opt. Instru. Eng., Bellingham, WA, 1991, p. 256.

Transmission Electron Microscopy (TEM)* 20

In transmission electron microscopy (TEM) a thin solid specimen (≤ 200 nm thick) is bombarded in vacuum with a highly-focused, monoenergetic beam of electrons. The beam is of sufficient energy to propagate through the specimen. A series of electromagnetic lenses then magnifies this transmitted electron signal. Diffracted electrons are observed in the form of a diffraction pattern beneath the specimen. This information is used to determine the atomic structure of the material in the sample. Transmitted electrons form images from small regions of sample that contain contrast, due to several scattering mechanisms associated with interactions between electrons and the atomic constituents of the sample. Analysis of transmitted electron images yields information both about atomic structure and about defects present in the material.

Range of elements	TEM does not specifically identify elements measured
Destructive	Yes, during specimen preparation
Chemical bonding information	Sometimes, indirectly from diffraction and image simulation
Quantification	Yes, atomic structures by diffraction; defect characterization by systematic image analysis
Accuracy	Lattice parameters to four significant figures using convergent beam diffraction
Detection limits	One monolayer for relatively high-Z materials
Depth resolution	None, except there are techniques that measure sample thickness
Lateral resolution	Better than 0.2 nm on some instruments
Imaging/mapping	Yes
Sample requirements	Solid conductors and coated insulators. Typically 3-mm diameter, < 200-nm thick in the center
Main uses	Atomic structure and microstructural analysis of solid materials, providing high lateral resolution
Instrument cost	\$300,000–\$1,500,000
Size	100 ft^2 to a major lab

Variable-Angle Spectroscopic Ellipsometry (VASE)* 21

In variable-angle spectroscopic ellipsometry (VASE), polarized light strikes a surface and the polarization of the reflected light is analyzed using a second polarizer. The light beam is highly collimated and monochromatic, and is incident on the material at an oblique angle. For each angle of incidence and wavelength, the reflected light intensity is measured as a function of polarization angle, allowing the important ellipsometric parameter to be determined. An optimum set of angle of incidence and wavelength combinations is used to maximize measurement sensitivity and information obtained. Physical quantities derivable from the measured parameter include the optical constants of bulk or filmed media, the thicknesses of films (from 1 to a few hundred nm), and the microstructural composition of a multiconstituent thin film. In general only materials with parallel interfaces, and with structural or chemical inhomogeneities on a scale less than about one-tenth the wavelength of the incident light, can be studied by ellipsometry.

Main use	Film thicknesses, microstructure, and optical properties
Optical range	Near ultraviolet to mid infrared
Sample requirements	Planar materials and interfaces
Destructive	No, operation in any transparent ambient, including vacuum, gases, liquids, and air
Depth probed	Light penetration of the material (tens of nm to μm)
Lateral resolution	mm normally, 100 μm under special conditions
Image/mapping	No
Instrument cost	\$50,000–\$150,000
Size	0.5 m × 1 m

X-Ray Diffraction (XRD)* 22

In X-ray diffraction (XRD) a collimated beam of X rays, with wavelength $\lambda \approx$ 0.5–2 Å, is incident on a specimen and is diffracted by the crystalline phases in the specimen according to Bragg's law ($\lambda = 2d \sin \theta$, where d is the spacing between atomic planes in the crystalline phase). The intensity of the diffracted X rays is measured as a function of the diffraction angle 2θ and the specimen's orientation. This diffraction pattern is used to identify the specimen's crystalline phases and to measure its structural properties, including strain (which is measured with great accuracy), epitaxy, and the size and orientation of crystallites (small crystalline regions). XRD can also determine concentration profiles, film thicknesses, and atomic arrangements in amorphous materials and multilayers. It also can characterize defects. To obtain this structural and physical information from thin films, XRD instruments and techniques are designed to maximize the diffracted X-ray intensities, since the diffracting power of thin films is small.

Range of elements	All, but not element specific. Low-Z elements may be difficult to detect
Probing depth	Typically a few µm but material dependent; monolayer sensitivity with synchrotron radiation
Detection limits	Material dependent, but ~3% in a two phase mixture; with synchrotron radiation can be ~0.1%
Destructive	No, for most materials
Depth profiling	Normally no; but this can be achieved.
Sample requirements	Any material, greater than ~0.5 cm, although smaller with microfocus
Lateral resolution	Normally none; although ~10 µm with microfocus
Main use	Identification of crystalline phases; determination of strain, and crystallite orientation and size; accurate determination of atomic arrangements
Specialized uses	Defect imaging and characterization; atomic arrangements in amorphous materials and multilayers; concentration profiles with depth; film thickness measurements
Instrument cost	\$70,000–\$200,000
Size	Varies with instrument, greater than ~70 ft^2

X-Ray Fluorescence (XRF)* 23

In X-ray fluorescence (XRF), an X-ray beam is used to irradiate a specimen, and the emitted fluorescent X rays are analyzed with a crystal spectrometer and scintillation or proportional counter. The fluorescent radiation normally is diffracted by a crystal at different angles to separate the X-ray wavelengths and therefore to identify the elements; concentrations are determined from the peak intensities. For thin films XRF intensity–composition–thickness equations derived from first principles are used for the precision determination of composition and thickness. This can be done also for each individual layer of multiple-layer films.

Range of elements	All but low-Z elements: hydrogen, helium, and lithium
Accuracy	±1% for composition, ±3% for thickness
Destructive	No
Depth sampled	Normally in the 10-μm range, but can be a few tens of Å in the total-reflection range
Depth profiling	Normally no, but possible using variable-incidence X rays
Detection limits	Normally 0.1% in concentration.
Sensitivity	10–10^5 Å in thickness can be examined
Lateral resolution	Normally none, but down to 10 μm using a microbeam
Chemical bond information	Normally no, but can be obtained from soft X-ray spectra
Sample requirements	≤5.0 cm in diameter
Main use	Identification of elements; determination of composition and thickness
Instrument cost	\$50,000–\$300,000
Size	5 ft × 8 ft

X-Ray Photoelectron Spectroscopy (XPS)* 24

In X-ray photoelectron spectroscopy (XPS), monoenergetic soft X rays bombard a sample material, causing electrons to be ejected. Identification of the elements present in the sample can be made directly from the kinetic energies of these ejected photoelectrons. On a finer scale it is also possible to identify the chemical state of the elements present from small variations in the determined kinetic energies. The relative concentrations of elements can be determined from the measured photoelectron intensities. For a solid, XPS probes 2–20 atomic layers deep, depending on the material, the energy of the photoelectron concerned, and the angle (with respect to the surface) of the measurement. The particular strengths of XPS are semiquantitative elemental analysis of surfaces without standards, and chemical state analysis, for materials as diverse as biological to metallurgical. XPS also is known as electron spectroscopy for chemical analysis (ESCA).

Range of elements	All except hydrogen and helium
Destructive	No, some beam damage to X-ray sensitive materials
Elemental analysis	Yes, semiquantitative without standards; quantitative with standards. Not a trace element method.
Chemical state information	Yes
Depth probed	5–50 Å
Depth profiling	Yes, over the top 50 Å; greater depths require sputter profiling
Depth resolution	A few to several tens of Å, depending on conditions
Lateral resolution	5 mm to 75 μm; down to 5 μm in special instruments
Sample requirements	All vacuum-compatible materials; flat samples best; size accepted depends on particular instrument
Main uses	Determinations of elemental and chemical state compositions in the top 30 Å
Instrument cost	$200,000–$1,000,000, depending on capabilities
Size	10 ft × 12 ft

Index